Robert Kemp Philp

The Reason Why

A Careful collection of many hundreds of reasons for things which though generally relieved are imperfectly understood

AF874267

Elibron Classics
www.elibron.com

Elibron Classics series.

ISBN 0-543-70287-1 (paperback)
ISBN 0-543-70286-3 (hardcover)

This Elibron Classics Replica Edition is an unabridged facsimile of the edition published in 1859 by Houlston and Wright, London.

THE REASON WHY.

THE

REASON WHY

A CAREFUL

COLLECTION OF MANY HUNDREDS OF REASONS FOR THINGS WHICH, THOUGH GENERALLY BELIEVED, ARE IMPERFECTLY UNDERSTOOD.

GENERAL SCIENCE.

"Stand still, and consider the wondrous works of God.

"Dost thou know when God disposed them, and caused the light of his cloud to shine?

"Dost thou know the balancings of the clouds, the wondrous works of him which is perfect in knowledge."—JOB XXXVII. 14—16.

THIRTY-FIRST THOUSAND.

LONDON
HOULSTON AND WRIGHT
65, PATERNOSTER ROW
MDCCCLIX.

1859.

HOULSTON AND WRIGHT'S PUBLICATIONS.

Crown 8vo., with numerous Illustrations, cloth, gilt edges, 3s. 6d.,

GREAT FACTS: a Popular History and Description of the most Remarkable Inventions during the present Century. By FREDERICK C. BAKEWELL, Author of "Philosophical Conversations," "Manual of Electricity," &c.

Demy 18mo., cloth, gilt edges, 3s. 6d.,

ENGLAND AND ITS PEOPLE; or, a Familiar History of the Social and Domestic Manners of its Inhabitants.

Square 16mo., cloth, gilt edges, 2s. 6d.,

DRAWING FOR YOUNG CHILDREN. Containing 150 Drawing Copies.

Square 16mo., cloth, gilt edges, 2s. 6d.,

EXERCISES IN LIGHT, SHADE, AND COLOUR, with numerous Coloured Diagrams.

Crown 8vo., with 300 explanatory Illustrations, cloth, 7s. 6d.,

THE USEFUL METALS AND THEIR ALLOYS, including MINING VENTILATION and JURISPRUDENCE, and the application of IRON, ZINC, TIN, &c., to the Industrial Arts. By the following AUTHORS :—

JOHN SCOFFERN, M.B., London.
WILLIAM FAIRBAIRN, F.R.S., Manchester.
WILLIAM TRURAN, C.E., Dowlais Works.
W. E. AITKIN, Birmingham.
WILLIAM CLAY, Esq., Mersey Works.
W. VOSE PICKETT, Esq.
ROBERT OXLAND, Plymouth.

Crown 8vo., cloth, 2s. 6d.,

THE DICTIONARY OF DAILY WANTS. Vol. I. By the Author of "Enquire Within." [*To be completed in 3 vols.*]

Crown 8vo., cloth limp, One Shilling,

THE YOUNG DEBATER: a Handbook for Mutual Improve-ment and Debating Societies. By SAMUEL NEIL, Author of "Composition and Elocution," "The Art of Reasoning," &c.

Second Edition, Crown 8vo., cloth limp, One Shilling,

COMPOSITION AND ELOCUTION; or, How to Speak and Write the English Language Correctly. By SAMUEL NEIL, Author of the "Young Debater," &c.

Crown 8vo., cloth limp, One Shilling,

THE USEFUL TEACHER, comprising English Grammar, His-tory, and Geography. By the Author of the "The Reason Why."

Each of the Treatises can be had separately, price 3d. sewed.

Crown 8vo., sewed, Sixpence,

CHAMPION'S ESSENTIALS OF THE FRENCH LANGUAGE; or, How to Read, Write, and Speak it Correctly.

Demy 8vo., sewed, One Shilling,

THE PROGRESS OF AGRICULTURE, from the Earliest Period down to the Meeting of the Agricultural Society at Chester, in August, 1858. By R. K. PHILP.

Demy 8vo., sewed, One Shilling,

THE PROGRESS OF CARRIAGES, ROADS, AND WATER Conveyances; from the Earliest Times to the Formation of Railways. By R. K. PHILP.

Crown 8vo., cloth limp, 1s. 6d.,

SOCIAL ETIQUETTE, THE ART OF COOKING, and Hints on Carving.

Crown 8vo., cloth limp, 1s. 6d.,

HOUSEHOLD MEDICINE, SURGERY, and DIET for Invalids.

HOULSTON & WRIGHT, 65, PATERNOSTER ROW, LONDON.

PREFACE.

We are all children of one Father, whose Works it should be our delight to study. As the intelligent child, standing by his parent's knee, asks explanations alike of the most simple phenomena, and of the most profound problems; so should man, turning to his Creator, continually ask for knowledge.

The greatest philosophers have been those who have clung to the demonstrative sciences, and have held that a simple truth well ascertained, is greater than the most ingenious theory founded upon questionable premises. Newton made more scientific revelations to mankind than any other philosopher; his discoveries have borne the searching test of time, because he snatched at nothing, leaped over no chasm to establish a favourite dogma; but, by the slowest steps, and by regarding the merest trifles, as well as the highest phenomena, he learnt to read Nature correctly. He discovered that her atoms of matter were letters, her blades of grass were words, her phenomena were sentences, and her complete volume a grand poem, teaching on every page the wisdom and the power of an Almighty Creator.

When he observed an apple fall to the ground, he asked the "reason why;" and in answer to that inquiry, there came one of the grandest discoveries that has ever been recorded upon the book of science. With that discovery a flood of light burst upon the human mind, illustrating in a far higher degree than had ever previously been conceived, the vastness of Almighty Power.

Every flower, every ray of light, every drop of dew, each flake of snow, the curling smoke, the lowering cloud, the bright sun, the pale moon, the twinkling stars, speak to us in eloquent language of the great Hand that made them. But millions lose the grand lesson which Nature teaches, because they can attach no meaning to what they see or hear.

The questions which appear in the succeeding pages are such as naturally occur to every inquiring mind; many of them are "every-day questions," frequently asked, but seldom answered satisfactorily, and, therefore, repeated again and again. No person can look through these pages without being reminded how frequently the same inquiry has occurred to him; every father can testify that children, from their earliest years, begin to ask the "reason why" for many things that are herein asked and explained.

In the compilation of such a work as this, there is no difficulty in the remembrance of such questions as are commonly heard, nor in the invention of such as, being less common, are necessary to the complete elucidation of a subject. The real, and the *only* difficulty, is to supply such answers as are clear, concise, and accurate; in fact, to satisfy the mind of the inquirer.

For the thousands of interesting truths which this volume contains, we are indebted to the labours of those great men who have devoted their lives to the solution of nature's problems, and have enabled us to give the "reason why" for many things upon which, but for their great genius and unremitting labours, we ourselves had remained in ignorance. Uninstructed by Newton and Huyghens, what could *we* have said upon the laws and properties of light?

"Nature's resplendent robe
Without whose vesting beauty all were wrapt
In unessential gloom."

We might easily have asked *why* the flower, the butterfly, or the bird, displayed such gorgeous hues; but could *we* have answered that in the magic beam itself lay all the glories of the many-tinted world? Untaught by Harvey, what could *we* have written of those wonderful rivers in the human frame through which the vital fluid runs? But for Wells, that earnest inquirer who, though sinking from a delicate constitution, went forth by night, in storm or starlight, in warmth or frost, and patiently watched the deposition of dew, zealously pursuing his observations through a series of years, what could *we* have said of those interesting laws that bespangle leaves and flowers with cooling drops, when, after the heat of a summer's day, they droop, and appear to ask for drink? We could have asked why the grass plot becomes wet with dew? and why the gravel walk remains comparatively dry? This were easy; but could we have told of those laws of radiation and condensation that fill the cup of the thirsty flower, but withhold the dew-drop from the lifeless stone?

To such learned and philanthropic men we are indebted for the philosophy of our "reasons:" men who gave to the world the great truths they had laboured to discover, and to the investigation of which they had devoted their lives, asking no other reward than that the knowledge they had gleaned might be freely communicated to mankind.

Among the various works that have aided the familiar exposition of science, "*Gower's Scientific Phenomena of Domestic Life*" is one of the most interesting. In "*Dr. Lardner's Treatise upon Heat,*" he also has adopted, in many parts, the most simple mode of illustration; and we beg specially to acknowledge the courtesy of Messrs. Longman and Co., the publishers, for their permission to adopt many of the familiar illustrations of science included in the works referred to. We further attach a list of the authorities consulted in the compilation of "THE REASON WHY," and to express our obligations to the men of science and letters who have aided us in our labours.

The volume having been carefully revised for the present edition, will retain its present arrangement, and will only undergo in future editions such verbal and statistical alterations as may be required to bring it up to the latest information upon scientific facts.

London, 1858.

LIST OF AUTHORITIES AND WORKS

CONSULTED IN THE

COMPILATION OF THE REASON WHY.

Accum's Theoretical and Practical Chemistry. *Kearsley.*
Arnott's Elements of Physics. *Longman & Co.*
Barrow's Popular Dictionary of Facts. *J. Hughes.*
Bernay's First Lines in Chemistry. *Parker.*
Bernay's Household Chemistry. *Parker.*
Bernan's History and Art of Warming, &c. *G. Bell.*
Black's Elements of Chemistry. *Longman & Co.*
Blair's First Catechisms. *Darton.*
Blair's Why and Because. *Darton & Co.*
Buff's, H., Familiar Letters. *Taylor & Walton.*
Chambers' Information for the People. *Chambers.*
Combe's Principles of Physiology. *Simpkin.*
Cooper's Light and Heat. *Grigg & Co., Philadelphia.*
Davidson's Treatise on Diet and Food. *Churchill.*
Davy, Sir H., on the Safety Lamp. *Weale.*
Dick's Regimen and Diet. *Whittaker.*
Foster's First Principles of Chemistry. *Harper, New York.*
Galloway's First Step in Chemistry. *Churchill.*
Gower's, C. F., Scientific Phenomena. *Longman & Co.*
Griffiths' Elements of Chemistry. *Parker.*
Griffiths' Chemistry of the Four Seasons. *Churchill.*
Griffin's Chemical Recreations. *Griffin.*
Heat, its Sources. Influences, and Results. *Tract Society.*
Howard's Theory of Clouds.
Hooker's Child's Book of Nature. *Low, Son, & Co.*
Hunt's Poetry of Science *Bohn.*
Humboldt's Aspects of Nature (translated by Mrs. Sabine). *Longman & Co.*
Humboldt's Cosmos, a Physical Description of the Universe (translated under the direction of Col. H. Sabine). *Longman & Co.*
Jamieson's Universal Science. *Whittaker.*
Johnston's Physical Atlasses. *Blackwood.*
Johnston's, James, Chemistry of Common Life. *Blackwood.*
Joyce's Outlines of Knowledge. *Lacey.*
Joyce's Scientific Dialogues. *Bohn.*
Joyce's Companion to dialogues. *Simpkin.*
Kenny's Philosophy Made Easy, or the Plain Why and Because. *Allman & Son.*
Knox's Manual of Anatomy. *Renshaw.*
Lardner's Treatise on Heat. *Longman.*
Leslie's Experimental Enquiries.
Lord's Popular Physiology. *Parker.*
Mitchell's First Lines of Science. *Tegg.*
Mitchell's outlines of Natural Philosophy. *Allman.*
Mudie's Spring, Summer, Autumn, and Winter. *Ward.*
Mudie's Air. *Ward.*
Mudie's Heavens. *Ward.*
Newth's First Book of Natural Philosophy. *Taylor & Co.*
Orr's Circle of the Sciences. *Houlston.*
Paley's Natural Theology, with Notes by Lord Brougham and Sir Charles Bell. *Knight.*
Parkes's Chemical Catechism. *Baldwin and Co.*
Phillip's, Sir R., Million of Facts. *Darton.*
Philosophy in Sport made Science in Earnest. *Murray.*
Pinnock's Catechisms. *Whittaker & Co.*
Potter's Elementary Optics. *Taylor & Co.*
Potter's Physical Optics. *Taylor & Co.*
Reid's H., Catechism of Heat. *Simpkin.*
Reid's Catechism of Chemistry. *Simpkin.*
Reid's, D., Ventilation. *Longman & Co.*
Smith's Philosophy of Health. *Knight.*
Somerville's Physical Geography. *Murray.*
Thomson's, R. D., School Chemistry. *Longman & Co.*
Thomson's, T., System of Chemistry. *Baldwin.*
Thomson's, T., Outlines of Science, Heat, and Electricity. *Baldwin.*
Thomson's R. D., Cyclopedia of Chemistry. *Longman & Co.*
Three Hundred Weather Predictions. *Bogue.*
Timb's Knowledge for the People, or the Plain Why and Because. *Low.*
Timb's Manuals of Utility. *Bogue.*
Timb's Things not Generally Known. *Bogue.*
Valentin's Text Book of Physiology (translated from the German, by W. Brinton, M.D.). *Renshaw.*
Well's Experimental Inquiry on Dew.
Wilson's Anatomists' Vade Mecum. *Churchill.*

INDEX, AND INDEX LESSONS.

☞ *The numbers refer to the Questions. The Index Lessons do not correspond with the Chapters, but are designed to bring together, in their alphabetical connection, all the Questions and Answers upon each particular subject included in the Work.*

LESSON I.

LESSON II.

LESSON III.

LESSON IV.

LESSON V.

LESSON VI.

LESSON VII.

LESSON VIII.

LESSON IX.

LESSON X.

LESSON XI.

LESSON XII.

LESSON XIII.

LESSON XIV.

LESSON XV.

LESSON XVI.

LESSON XVII.

LESSON XVIII.

LESSON XIX.

LESSON XX.

LESSON XXI.

LESSON XXII.

LESSON XXIII.

LESSON XXIV.

LESSON XXV.

LESSON XXVI.

LESSON XXVII.

LESSON XXVIII.

LESSON XXIX.

LESSON XXX.

LESSON XXXI.

LESSON XXXII.

LESSON XXXIII.

LESSON XXXIV.

LESSON XXXV.

LESSON XXXVI.

LESSON XXXVII.

LESSON XXXVIII.

LESSON XXXIX.

LESSON XL.

LESSON XLI.

LESSON XLII.

LESSON XLIII.

LESSON XLIV.

LESSON XLV.

LESSON XLVI.

LESSON XLVII.

LESSON XLVIII.

LESSON XLIX.

LESSON L.

LESSON LI.

LESSON LII.

LESSON LIII.

LESSON LIV.

LESSON LV.

LESSON LVI.

LESSON LVII.

LESSON LVIII.

LESSON LIX.

LESSON LX.

LESSON LXI.

LESSON LXII.

LESSON LXIII.

LESSON LXIV.

LESSON LXV.

LESSON LXVI.

LESSON LXVII.

LESSON LXVIII.

LESSON LXIX.

LESSON LXX.

LESSON LXXI.

LESSON LXXII.

LESSON LXXIII.

LESSON LXXIV.

LESSON LXXV.

LESSON LXXVI.

LESSON LXXVII.

LESSON LXXVIII.

LESSON LXXIX.

LESSON LXXX.

"God looked down from heaven upon the children of men, to see if there were any that did understand, that did see God."—PSALM LIII.

THE REASON WHY.

CHAPTER I.

1. *Why should we seek knowledge?*

Because it assists us to comprehend the *goodness and power of God;*

And gives us power over the circumstances and associations by which we are surrounded: the proper exercise of this power will greatly promote our happiness.

2. *Why does the possession of knowledge enable us to exercise power over surrounding circumstances?*

Knowledge enables us to understand that to live healthily we require to breathe fresh and pure air. It also tells us that animal and vegetable substances undergoing decay poison the air, though we may not be able to see, or to smell, or otherwise discover the existence of the poison. Knowing this, we become careful to remove from our presence all matters that tend to corrupt the atmosphere. This is only one of the countless instances in which knowledge gives us power over surrounding circumstances.

3. *Name some other instances in which knowledge gives us power.*

Knowledge of *Geography* and *Navigation* enables the mariner to guide his ship across the trackless deep, and to reach the sought-for port, though he had never before been on its shores.

Knowledge of *Chemistry* enables us to separate or to combine the various substances found in nature. Thus we obtain useful and precious metals from what at first appeared to be worthless stones transparent glass from pebbles, through which no light could pass; soap from oily substances; and gas from solid bodies. Matter, which has lain for ages in the depths of the earth, is disentombed, and made to warm and illuminate our homes. Though it has long been buried in the darkness of the grave, it is made to impart light and cheerfulness to millions who live. The rough and opaque stone is melted into a transparent sheet, which, being placed in our windows, keeps out the cold blasts of winter, and turns away the soaking showers of rain, or the chilling flakes of snow, and stones of hail; but, like a fairy curtain, the crystal sheet allows light and warmth to enter, carrying comfort to the hearts, and beauty to the eyes, of those who dwell within.

Knowledge of *Medicine* enables the physician to overcome the ravages of disease, and to save suffering patients from sinking prematurely to the grave.

Knowledge of *Anatomy* and *Surgery* enables the surgeon to bind up dangerous fractures and wounds, and to remove, even from the internal parts of bodies, ulcers and diseased formations that would otherwise be fatal to life.

Knowledge of *Mechanics* enables man to increase his power by the construction of machines. The steam-ship crossing the ocean in opposition to wind and tide, the railway locomotive travelling at sixty miles an hour, and the steam-hammer beating blocks of iron into useful shapes, are evidences of the power which man acquires through knowledge of mechanics.

Knowledge of *Electricity* enables man to stand in comparative safety amid the awful war of elements. Lightning, the offspring of electricity, has a tendency to strike upon lofty objects by which it may be attracted. By its mighty powers churches or houses may be instantly levelled with the dust. But man, knowing that electricity is strongly attracted by particular substances, raises over lofty buildings rods of metal communicating with bars that descend into the ground. The lightning, rushing with indescribable force toward the steeple, is attracted by the bars of metal, and conducted harmlessly to the earth. Man may thus be said to take even lightning by the hand, and to divert its destroying force by the aid of Knowledge. And in countless other instances "Knowledge is Power."

"And God said, Let there be light: and there was light. And God saw the light, that it was good: and God divided the light from the darkness."—GEN. I.

CHAPTER II.

4. *What is light?*

Light, according to Newton, is the effect of luminous particles of ether which dart from the surfaces of bodies in all directions. According to his theory, the solar light which we receive *departs from the sun and travels to the earth.*

According to Huyghens, light is caused by an *infinitely elastic ether, diffused through all space.* This ether, existing everywhere, is *excited into waves, or vibrations, by the luminous body.*

5. The theory of light is undetermined, so that neither the views of Newton, nor those of Huyghens, can be said to be exclusively adopted. Writers upon natural philosophy seize hold of either or both of those theories, as they present themselves more or less favourably in the explanation of natural phenomena. In *The Reason Why*, as we have to speak of the *effects* of light rather than of its *cause*, we shall state the two theories, but avoid discussion upon them. Let no one be discouraged by the fact that the theory of light, as, indeed, of all the imponderable agents, is imperfectly understood. Rather let us rejoice that there are vast fields of discovery to be explored; and that light, the most glorious and inspiring element in nature, invites us from the sun, the moon, and the stars, and from the face of every green leaf and variegated flower, to search out the wonders of its nature, and to exemplify the goodness and wisdom of God.

6. *What is heat, otherwise called caloric?*

Heat is a principle in nature which, like light and electricity, is best understood by its *effects*. We popularly call that heat which raises the temperature of bodies submitted to its influence.

Caloric is another term for heat. It is advisable, however, to use the term *caloric* when speaking of the *cause* of heat, and of *heat* as the *effect* of the presence of *caloric*.

Heat may exist without *fire* or *light*.
It is not perceptible to *vision*.
It makes an impression upon our *feelings*.
It acts powerfully upon *all bodies*.
It has no *weight*.
It attends, or is connected with, *all the operations of nature*.
It radiates from *all bodies* in straight lines, and in all *directions*.
It strikes most powerfully in *direct lines*.

Its rays may be collected into a *focus*, just as the rays of the sun.
It may be *reflected* from a polished surface.
It is more easily *conducted* by some substances than by others.

7. *What is electricity?*

Electricity is a property of *force* which resides in all matter, and which constantly seeks to establish an *equilibrium*.

8. *Why is it called electricity?*

Because it first revealed itself to human observation through a substance called, in the Greek language, *electrum*. This substance is known to us as *amber*.

9. *What is amber?*

It is a *resinous* substance, hard, bitter, shapeless, and glossy. It has been variously supposed to be a vegetable gum, a fossil, and an animal product. It is probably formed by a *species of ant* that inhabit pine forests. The bodies of ants are frequently found in its substance.

10. *In what way did electrum induce attention to this property of force in matter?*

Thales, a Greek philosopher, observed that, by briskly rubbing *electrum*, it acquired the property of *attracting* light particles of matter, which moved towards the amber, and attached themselves to its surface, evidently under the influence of a *force* excited in the amber.

11. *What is magnetism?*

Magnetism is *the electricity of the earth*; it is characterised by the circulation of *currents of electricity passing through the earth's surface.*

12. *Why are light, heat, and electricity called imponderable elements?*

Because, though they possess decided properties, and produce most remarkable effects, they differ from other elements in being destitute of *weight*. A substance becomes no heavier by being made *hot*; in fact, it becomes specifically lighter. An intense light, though it compels us to close our eyes, has *no weight;* and bodies, when charged with electricity, are *no heavier* than when the electricity is withdrawn.

Imponderable—not having sensible weight. *Ponderable*—capable of being weighed.

13. *What is matter?*

The term matter is employed to designate the substances of which all created things are formed; but though we understand many of the properties and laws of matter, we do not know *what matter is.*

14. *What are the distinguishing states of matter?*

The *imponderable*—as that of light, heat, and electricity; the *aeriform*—as that of gas or air; the *liquid*—as that of water or spirit; and the *solid*—as that of stone or iron.

15. *What are the recognised elements of matter?*

The recognised *elements* of matter are those portions of it of which the properties are supposed to be known, and which enter into various combinations with each other to produce all the varied substances that exist.

16. *How many elements of matter are now known?*

There are sixty-two. But it is not improbable that science, in its progress, will discover more.

17. *Are light, heat, and electricity included in the sixty-two elements?*

No; the sixty-two are all of them known to be *substances,* to possess *weight.* They are called *ponderable* elements, in distinction from the *imponderable*

There are various opinions as to whether light, heat, and electricity are *substances;* and the terms *ether*, *fluid*, &c., are employed as the best expressions of our imperfect ideas respecting them.

18. *What is understood to be the constitution of the elements of matter?*

The *elements of matter* of every kind are understood to consist of *atoms,* too minute to be observed; and the *properties of bodies*

"The day is thine, the night also is thine: thou hast prepared the light and the sun."—PSALM LXXIV.

are believed to be modified by the *nature* and the *number* of their ultimate atoms, and the extent to which *light, heat,* and *electricity* may participate in their union.

19. *What are the phenomena of matter?*

The *phenomena of matter* include all the various and observable states and changes which nature undergoes. To understand the phenomena of nature is to be able to read her wonders, comprehend her beauties, and perceive in all things that exist the hand of a wise Creator. The most interesting and instructive phenomena are momentarily occurring in our dwellings, our gardens, and surround us in our wanderings. The phenomena that attend the evaporation of steam from a kettle, are as interesting as those which accompany the formation of clouds in the far regions of the atmosphere; the phenomena displayed by the frost upon the window pane, as illustrative of grand natural laws as those which mark the sealing of the poles of the earth in mountains of ice.

Phenomenon—whatever is capable of observation; plural, *phenomena.*

CHAPTER III.

20. *What is the constitution of a ray of light?*

A ray of *white light,* as we receive it from the sun, is composed of *a number of elementary rays,* which, with the aid of a *prism,* may be separated, and will produce under refraction the following colours:—

1. An *extreme red* ray—a mixture of *red* and *blue* the red *predominating.*
2. *Red.*
3. *Orange*—red passing into and combining with yellow.
4. *Yellow*—the most luminous of all the rays.
5. *Green*—yellow passing into and combining with the blue.
6. *Blue.*
7. *Indigo*—a dark and intense blue.
8. *Violet*—blue mingled with red.
9. *Lavender grey*—a neutral tint.
10. Rays called *fluorescent,* which are either of a *pure silvery blue,* or a *delicate green.*

"Ye are the light of the world. A city that is set on a hill cannot be hid."—
MATT. V.

21. *At what rate of velocity does light travel?*

At the rate of 192,000 miles in a *second* through our *atmosphere;* and 192,500 miles in a *second* through a *vacuum.*

22. *How long does light take to travel from the sun to the earth?*

Eight minutes and thirteen seconds.

23. *What is the constitution of the sun?*

It is a spherical body, 1,384,472 times larger than the earth.

24. *From what does the luminosity of the sun arise?*

From a luminous atmosphere, or, as M. Arago named it, *photosphere,* which completely surrounds the body of the sun, and which probably *burns with great intensity.*

25. *Why may spots be observed upon the face of the sun. by the aid of a telescope?*

Because the *photosphere* which surrounds the sun is subject to violent disturbances (as is the case with most burning bodies); these disturbances give rise to openings through the luminous envelope of the sun, by which we are enabled to look in upon *the dark body of the sun* itself.

26. *What are the minor sources of light?*

Light may be produced by *chemical action,* by *electricity,* and by *phosphoresence,* in the latter of which various agencies unite.

27. *What is a ray of light?*

A *ray* of light is the *smallest portion* of light which we can recognise.

28. *What is a medium?*

A *medium* is a body which affords *a passage for the rays* of light.

29. *What is a beam of light?*

A *beam* of light is a *group of parallel rays.*

30. *What is a pencil of light?*

A *pencil* of light is a body of rays which *come from or move towards a point.*

"Light is sown for the righteous, and gladness for the upright in heart."—
PSALM XCVII.

31. *What is the radiant point?*

The *radiant point* is that *from which diverging rays of light are emitted.*

Diverging—starting from a point, and separating.

32. *What is the focus?*

The *focus* is the point to which *converging rays are directed.*

Converging—drawing together towards a point.

33. *Why are there so many varieties of colour and tint in the various objects in nature?*

Because every surface has a peculiar constitution, or atomic condition, *by which the light falling upon it is influenced.* In tropical climates, where the brightness of the sun is the most intense, there the colours of natural objects are the richest; the foliage is of the darkest green; the flowers and fruits present the brightest hues; and the plumage of the birds is of the most gaudy description. In the temperate climates these features are more subdued, still bearing relation to the degree of light. And at a certain depth of the ocean, where light penetrates only in a slight degree, the objects that abound are nearly colourless.

34. It has been held by many philosophers (and the theory is so far conclusive that it cannot be dispensed with) that there is an analogy between the vibratory causes of *sound*, and the vibratory causes of *colour*. Any one who has seen an Æolian harp, and listened to the wild notes of its music, will be aware that the wires of the harp are swept by accidental currents of air; that when those currents have been strong, the notes of the harp have been raised to the highest pitch, and as the intensity of the currents has fallen, the musical sounds have deepened and softened, until, with melodious sighing, they have died away. No finger has touched the strings; no musical genius has presided at the harp to wake its inspiring sounds; but the vibration imparted to the air, as it swept the wires, has alone produced the chromatic sounds that have charmed the listener. If, then, the varied vibrations of the *air* are capable of imparting dissimilar sensations of *sounds* to the *ear*, is it not only possible, but probable, that the different vibrations of *light* may impart the various sensations of *colours* to the *eye?*

35. *Why is a substance red, according to Newton's theory of light?*

Because it *absorbs* all but the red rays.

36. *Why is a substance red, according to Huyghen's theory?*

Because the peculiar *atomic condition* of its surface alters the *rapidity of the vibrations* of the ether producing light; and the altered number of vibrations produces *a sensation upon the eye* which we denominate a *red colour*.

The effect may be compared to an *echo* which *alters the tone of sound.* Light is *echoed* from the petals of a flower, and the tone of the light is affected by the nature of the surface from which it is reflected?

37. *Why is a substance yellow, according to Newton?*

Because it absorbs all but the *yellow rays.*

38. *Why is a substance yellow, according to Huyghens?*

Because it returns the light-producing ether vibrating at the rate of *five hundred and thirty-five billions* of times in a second.

39. *Why is a substance blue, according to Huyghens?*

Because it increases the number of vibrations *to six hundred and twenty-two billions* in a second.

40. *Why is a substance white, according to Newton?*

Because it returns *all* the component rays of light.

41. *Why is a substance white, according to Huyghens?*

Because it returns the luminous ether *without altering its vibrations.*

42. *Why is a substance black, according to Newton?*

Because it *absorbs* all the rays of light.

43. *Why is a substance black, according to Huyghens?*

Because it *puts an end to the vibrations* of the luminous ether.

44. Accepting the theory of vibrations, and applying it to the elucidation of the phenomena of light—it is unnecessary, we think, to believe that a ray of *white* light *contains* rays in *a state of colour*. It is said that if we divide a circular surface into parts, and paint the various colours in the order and proportions in which they occur in the refracted ray, and then spin the circle with great velocity, the colours will blend and appear *white*. But such is not the case; the result is in some degree an illusion, arising out of the sudden removal of the impression made upon the eye by the colours; and if a piece of white paper be held by the side of the coloured circle in motion, the latter will be found to be *grey*. When it is remembered that in colouring a white surface with thin colours, the white materially qualifies the colours, it must be admitted that the

experiment fails to support the assertion that the colours of the spectrum produce white. But there can be no difficulty in understanding that a ray of light undergoing *refraction* becomes divided into minor rays, which *differing in their degrees of refrangibility*, vary also in the *velocity of their vibrations*, and produce the several sensations of colour.

45. *When light is absorbed, what becomes of it?*

No satisfactory explanation can be given. It is generally believed that absorbed light is retained in the substances of the absorbing bodies *in a latent or hidden state.* And it has been supposed that the *light of electricity* may arise from the setting free of light thus absorbed and hidden.

46. *Why are some bodies transparent?*

Because their atoms are so disposed that they *do not interrupt the vibrations* of the light-giving ether.

47. *Are any bodies absolutely transparent?*

Not absolutely so; whatever medium light passes through, *some portion* of it is diverted. Even in traversing our atmosphere, some light is lost.

48. *Do both the theories of light admit of the existence of an ether in a state of vibration?*

Yes; in both theories *a wave motion of the luminous essence is admitted.*

The chief difference in the two theories is, that according to Newton, the luminous ether *comes from the sun*, and is reflected, refracted, or absorbed, by various bodies; while, according to Huyghens, the ether which produces light pervades all space, and is *excited to luminosity* by certain bodies, of which the sun is the chief.

Another difference is, that according to Newton the various colours of bodies arise from the *absorption* of some of the rays, and the *reflection* of others; and, in some cases, from the *refraction*, or bending, of the rays, by which their elementary colours are displayed. According to Huyghens, the colours of bodies arise from the *vibratory impressions* they impart to the luminous essence.

"There is no darkness nor shadow of death, where the workers of iniquity may hide themselves."—JOB XXXIV.

49. *What are the chief sources of light?*

Light is emitted from the *sun* and the *fixed stars;* and it is evolved by *percussion*, *electricity*, *combustion*, and by various *chemical processes.*

50. *Why is light evolved by the striking of a flint against steel?*

It is believed by some philosophers that light, when absorbed by bodies, becomes latent, *or hidden in them*, and that it is *set free* by the actions that cause it to re-appear. This opinion, which is full cf interest, opens a vast field of inquiry; but it is at present purely problematical. It applies equally to the light of *fires*, *candles*, and of all other minor sources. Recent photographic experiments sustain this theory.

51. *How long does light occupy in its flight from the sun to the earth?*

Light travels from the sun to the earth (a distance of ninety-five millions of miles) in a little more than *eight minutes.*

52. *How long would it take a cannon ball to reach the sun?*

A cannon ball, flying at the rate of about *a mile in eight seconds*, would occupy *thirty-two years* in going over the sun's distance from the earth. Light travels *two million times faster* than a cannon ball.

53. *Why do we see various shadows upon the ground, the walls, &c.*

Because light travels only *in straight lines.* Whenever, therefore, an opaque body intercepts a beam of light, the shadow of that body will be bounded by right lines, passing from the luminous body, and meeting the lines that terminate the opaque body.

54. *What is the reflection of light?*

The reflection of light is the *turning back* of its rays by the surface upon which they fall.

"I am come a light into the world, that whosoever believeth in me should not abide in darkness."—JOHN XIII.

55. *Why do bright surfaces, such as those of mirrors, metals, &c., reflect the images of objects?*

Because the rays of light which are sent to those surfaces by the objects are *reflected*, and as these rays retain the colours imparted to them by the bodies from which they were originally directed, they present to the eye perfect images of those bodies.

56. *Why cannot we see ourselves in a piece of plain glass as well as in a mirror silvered at the back?*

Because, when a ray of light enters a transparent medium, it continues to move on until it meets with an opaque surface; it then becomes reflected, or bent back. When the light meets with the opaque surface of the quicksilver, it turns back, *and passes out again at the same surface at which it entered.*

57. The parts of a mirror from which the quicksilver is rubbed away give no reflection that could assist the reflecting power of the quicksilver. That the surface of the glass does not reflect the image, is shown by the fact, that if you put the point of any object against the glass, the thickness between the point and the place where the reflection of it begins, will *show the exact thickness of the glass.*

58. *Why can we cast the light of the sun to a great distance by means of a piece of looking-glass?*

Because the piece of looking-glass *turns back* the rays of the sun, and directs them to any place forming an angle with a line perpendicular to the surface of the glass.

59. *What is the line of incidence?*

If a person stands in a direct line before a mirror, the line through which the light travels from him to the mirror is *the line of incidence.*

Incidence—falling on.

60. *What is the line of reflection?*

The *line of reflection* is the line in which the rays of light are returned from the image formed in the glass to the eye of the observer.

Reflection—a turning back.

"When I consider thy heavens, the work of thy fingers, the moon and the stars which thou hast ordained,

61. *What is the angle of incidence?*

The *angle* of incidence is the angle which rays of light, falling on a reflecting surface, make with a line perpendicular to that surface.

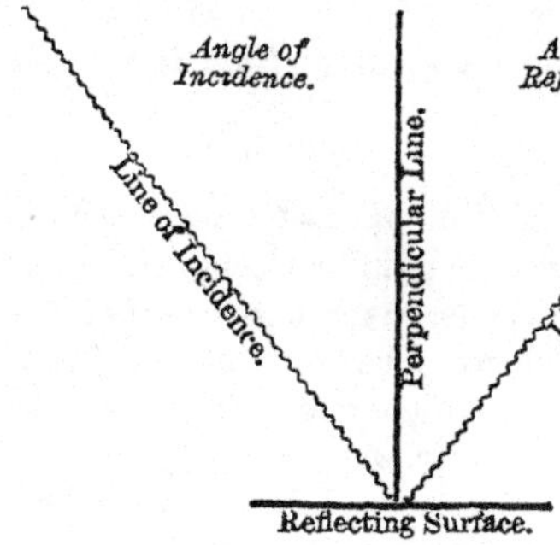

EXPLAINING THE LINES AND ANGLES OF INCIDENCE AND OF REFLECTION.

62. *What is the angle of reflection?*

The *angle* of reflection is the angle which is formed by the returning rays of light, and a line perpendicular to the reflecting surface. It is always *equivalent* to the angle of incidence.

63. Take a marble and roll it across the floor, so that it shall strike the wainscot obliquely. Let A in the diagram represent the point from which the marble is sent. The marble will not return to the hand, nor will it travel to the line B, but will bound off, or be *reflected*, to C. Now B is an imaginary line, *perpendicular to the reflecting surface;* and it will be found that the path described by the marble in *rolling to the surface and rebounding from it,* form, with the line B, two angles that are *equal.* These represent the angles of *incidence* and of *reflection,* and explain why the reflection of a person standing at A before a mirror, would be seen by another person standing at C. This simple law in optics explains a great many interesting phenomena, and therefore it should be clearly impressed upon the memory.

64. *Why does the imponderability of light show the goodness and wisdom of the Creator?*

Because, if light possessed weight, although its particles were equal only to the two millioneth part of a grain of sand, in consequence of *the wonderful velocity* of their flight, they would strike us with a force equal to that of sand when *shot point blank from a cannon's mouth.* This force would destroy every living thing.

"What is man, that thou art mindful of him? and the son of man, that thou visitest him?"—PSALM VIII.

65. *What is the refraction of light?*

When rays of light fall *obliquely* upon the surface of any *transparent medium*, they are slightly diverted from their course. This alteration of the course of the rays is called *refraction*, and the degree of refraction is influenced by the difference between the *densities* of the mediums *through which light is transmitted.*

66. *What is a prism?*

A prism is a *triangular piece of glass*, which is used in experiments respecting the nature of light and colours. In consequence of its varying thickness, it bends, or refracts, the rays of light passing through it. Colours which were before blended in the white ray (according to the Newtonian theory) are separated, because they differ in their degrees of refrangibility; they, therefore, divide, and exhibit what are called the *prismatic colours.* (See 20).

67. *Does a ray of light always display colours when it is refracted?*

No; the colours of the primitive rays do not appear, unless the degree of refraction is sufficient to *separate* the elementary rays.

68. *If a ray of light falls in a straight line upon a transparent surface, is it then refracted?*

In that case the ray pursues its course—*there is no refraction.*

69. *Is the direction in which the rays are bent, or refracted, influenced by the relative densities of the media?*

A ray of light falling slantingly upon a *window*, in passing through it is slightly brought to the *perpendicular;* and if it then falls upon the surface of water, it is still further brought to the perpendicular in *passing through the water.*

70. *Is light refracted in passing from a dense medium to a thinner one?*

It is; but the *direction of the refraction* is just the opposite to the instance just given; a ray of light passing through *water* into *air*, does not take a more *perpendicular course*, but becomes more *oblique*.

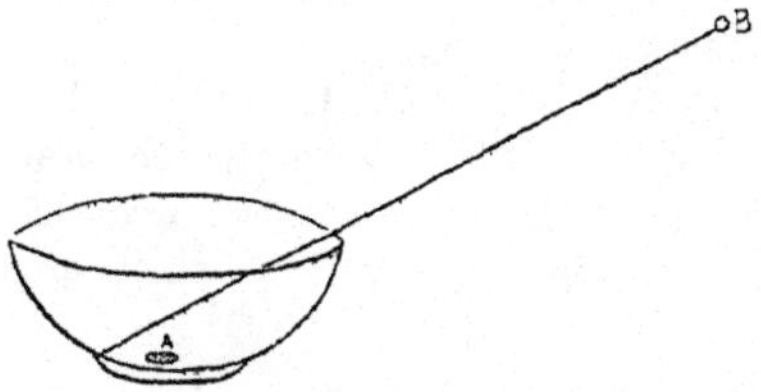

Fig. 1.—SHILLING AND BASIN WITHOUT WATER; SHOWING THE DIRECTION OF A RAY OF LIGHT.

71. Place in the bottom of an empty basin (Fig. 1) a shilling; then stand in such a position at the point B that the line of sight, over the edge of the basin, just excludes the shilling from view. Then request some one to pour water into the basin, until it is filled to C (Fig. 2), keeping your eye fixed upon the spot. The shilling will gradually appear, and will soon come entirely in view. Not only will the shilling be brought in view, but also portions of the basin before concealed. This is owing to the rays of light passing from the bottom through the water in a direction *more perpendicular* than they would have done through the air; but on leaving the water they become *more oblique*, and hence they convey the image of the shilling *over the edge of the basin*, which otherwise would have obstructed the view.

Fig. 2.—SHOWING THE BENDING OF THE RAY BY THE WATER.

72. *Why is it that in cloudy and showery days we see the sun's rays bursting through the clouds in different directions?*

Because, in passing through clouds of *different densities* the rays are *bent out of their course*.

"The earth is the Lord's, and the fulness thereof; the world, and they that dwell therein."—PSALM XXIV.

CHAPTER IV.

73. *What is the difference between light and heat?*

The most obvious distinction is, that light acts upon *vision*, and heat upon *sensation*, or feeling.

Another distinction is, that *heat expands all bodies*, and alters their atomic condition; while *light*, though usually attended by heat, does not display the same expansive force, but produces various effects which are *peculiar to itself*, and which are attributed to its *actinic* power.

74. *Are light and heat combined in the solar ray?*

Yes. A ray of light, as well as containing elementary rays that produce colours under refraction, contains also *chemical rays* and *heat rays*.

75. *How do we know that light and heat are separate elements?*

Because we have *heat rays*, as from dark hot iron, from various chemical actions, and from friction, which are *unattended by the development of light*. And we have light, or luminosity, such as that of *phosphoresence*, which is unaccompanied by any appreciable degree of heat.

But, besides this confirmation, further proof is afforded by the fact, that in passing rays of solar light through media that are *transparent to heat*, but not to *light*, the heat rays may be *separated* from the luminous rays, and *vice versa*.

76. Black glass, and black mica, which are nearly *opaque to light*, are *transparent to heat* to the extent of ninety degrees out of a hundred. While pale green glass, coloured by oxide of copper, and covered with a coating of water, or a thin coating of alum, will be perfectly *transparent to light*, but will be almost quite *opaque to heat*. These remarks apply, in a greater or less degree, to many other substances.

77. *In what respects are light and heat similar?*

Both heat and light have been referred to minute vibratory motions which occur, under exciting causes, in a very subtile elastic medium.

They are both united in the sun's rays.

They are both subject to laws of absorption, radiation, reflection, and refraction.

They are both essential to life, whether animal or vegetable.

Both may be developed in their greatest intensity by electricity.

They are both imponderable.

78. *In what respects are light and heat dissimilar?*

Heat frequently exists without light.

Light is usually attended with heat.

Light may be instantly extinguished, but

Heat can only be more gradually reduced by diffusion.

The solar rays deliver heat to the earth by day, and the heat remains with the earth when the light has departed.

Heat diffuses itself in all directions.

Light travels only in straight lines.

The colours that absorb and radiate both light and heat do not act in the same degree upon them both. Black, which does not radiate light, is a good *radiator of heat*, &c. &c.

The oxy-hydrogen *light* emits a most intense heat, but glass, which will transmit the rays of light, will afford no passage to the rays of the *heat*.

Heat is latent, and known to be in all bodies, but the theory of latent light is at present imperfectly understood.

These are only a few of the analogies and distinctions that exist between the two mysterious agents, light and heat. But they are sufficient to supply the starting points of investigation.

79. The importance of the heat that attends the solar rays may be illustrated by the experiments performed a few years ago, by Mr. Baker, of Fleet-street, London, who made a large burning lens, three feet and a half in diameter, and employed another lens to reduce the rays of the first to a focus of half an inch in diameter. The heat produced was so great that iron plates, gold, and stones were *instantly melted;* and sulphur, pitch, and resinous bodies, *were melted under water.*

80. *What is the relative intensity of primary and reflected light?*

The intensity of a reflection depends upon the power of the reflecting surface. But, taking the sun and moon as the great examples of primary and reflected light, the intensity of the *sun's light* is 801,072 times *greater than that of the moon.*

"Such knowledge is too wonderful for me; it is high, I cannot attain unto it."—PSALM CXXXIX.

81. *How do we know that objects reflect light in every direction?*

Because if we *prick a hole in a card with a pin,* and then look through that small hole upon a *landscape,* we can see some miles of country, and some thousands of objects; each part of every object throughout the whole scene must have sent rays of light to the small hole pricked in the card.

82. At one extremity of the landscape, viewed through the hole in the card, there may be a forest of trees; in the distance there may be hills bathed in golden light, and overhung with glittering clouds; in the mid-distance there may be a river winding its course along, as though it loved the earth through which it ran, and wished, by wandering to and fro, to refresh the thirsty soil; in the foreground may be a church, covered by a million ivy leaves; and grouping towards the sacred edifice may be hundreds of intending worshippers, old and young, rich and poor, flowers may adorn the pathways, and butterflies spangle the air with their beauties: yet every one of those objects—the forest, the hills, the clouds, the river, the church, the ivy, the people, the flowers, the butterflies—must have sent rays of light, which found their way through the little hole in the card, and entered to paint the picture upon the curtain of the eye.

This is one of the most striking instances that can be afforded of the wonderful properties of light, and of the infinitude of those luminous rays that attend the majestic rising of the sun. Not only does light fly from the grand "ruler of the day" with a velocity which is a million and a half times greater than that of a cannon-ball, but it darts from every reflecting surface with a like velocity, and reaches the tender structure of the eye so gently that, as it falls upon the little curtain of nerves which is there spread to receive it, it imparts the most pleasing sensations, and tells its story of the outer world with a minuteness of detail and a holiness of truth. Some philosophers once sought to *weigh the sunbeam;* they constructed a most delicate balance, and suddenly let in upon it a beam of light; the lever of the balance was so delicately hung that the fluttering of a fly would have disturbed it. Everything prepared, the grave men took their places, and with keen eyes watched the result. The sunbeam that was to decide the experiment had left the sun eight minutes prior to pass the ordeal. It had flown through ninety-five millions of miles of space in that short measure of time, and it shot upon the balance with unabated velocity, but the lever moved not, and the philosophers were mute!

83. *What is polarized light?*

Polarized light is light which has been subjected to *compound refraction,* and which, after polarization, exhibits a new series of phenomena, differing materially from those that pertain to the primary conditions of light.

"But if thine eye be evil, thy whole body shall be full of darkness. If therefore the light that is in thee be darkness, how great is that darkness."—MATT. v.

84. *What are the chief deductions from the phenomena observed under the polarization of light?*

The polarization of light appears to confirm in a high degree the vibratory *theory of light;* and to show that the vibrations of light have two planes or directions of motion. The mast of a ship, for instance, has two motions: it progresses *vertically* as the ship is impelled forward, and it rolls *laterally* through the motion of the billows.

Something like this occurs in the vibrations of light, only the *vertical vibration* is the condition of *one ray*, and the *lateral vibration* is the condition of another ray, and the vibrations of these two rays intersect each other in the solar ray. When these vibrations occur together, the ray has certain properties and powers. But by polarization the rays may be *separated*, and the result is two distinct rays, having *different vibrations.*

It then appears that various bodies are transparent to these polarized rays *only in certain directions.* And this fact is supposed to show that bodies are made up of their atoms arranged in certain planes, through or between which the *lateral* or the *vertical* waves of light, together or singly, can or cannot pass; and that the transparency or the opacity of a body is determined by the *relation of its atomic planes* to *the planes of the vibrations of light.*

Ordinary light, passing through transparent media, produces no very remarkable effect in its course; but *polarized light* appears to illuminate every atom of the permeated substance, and by surrounding it with a prismatic clothing, to afford an illustration of its *molecular arrangement.*

85. *Why are two persons able to see each other?*

Because rays of light *flow from their bodies to each other's eyes*, and convey an impression of their respective conditions.

86. In some popular works that have come under our notice, we find that the student is told that "we cannot absolutely see each other—we only *see the rays of light reflected from each other.*" The statement is erroneous as expressed. We do not see the *rays* of light, for if we did so, the effect of vision would be destroyed, and all bodies would *appear* to be in a state of *incandesence*, or of *phosphoresence.* Rays of light, which are themselves *invisible*, radiate from

"The hay appeareth, and the tender grass showeth itself, and herbs of the mountain are gathered."—PROV. XXVII.

the objects we look upon, enter the pupil of the eye, and impress the seat of vision in a manner which conveys to the mind a knowledge of the form, colour, and relative size and position of the figure we look upon. If this is not seeing the object—*what is?* It would be just as reasonable to say, that we cannot *hear* a person speak—that we only hear the *vibrations of the air*. But as the vibrations are imparted to the air by the organs of voice of the speaker, as he sets the air in motion, and makes the air his messenger to us, we certainly hear *him*, and can dispense with any logical myths that confound the understanding, and contribute to no good result.

CHAPTER V.

87. *What is actinism?*

Actinism is the chemical property of light.

Actinism—ray power.

88. *Why does silver tarnish when exposed to light?*

Because of the *actinic*, or chemical power of the rays of the sun.

89. *Why do some colours fade, and others darken, when exposed to the sun?*

Because of the *chemical* power of the sun's rays.

90. *Why can pictures be taken by the sun's rays?*

Because of the actinic powers that accompany the solar light.

91. *What is the particular chemical effect of light exhibited in the production of photographic pictures?*

Simply the *darkening of preparations of silver, by the actinic rays.*

92. *Why are photographic studios usually glazed with blue glass?*

Because blue glass obstructs many of the luminous rays, but it is perfectly transparent to *actinism*.

93. *Why do plants become scorched under the unclouded sun?*

Because the heat rays are in excess. The clouds shut off the scorching light; but, like the blue glass of the photographer's studio, they transmit *actinism*.

"While the earth remaineth, seed-time and harvest, and cold and heat, and summer and winter, and day and night, shall not cease."—GEN. VIII.

94. *What effect has actinism upon vegetation?*

It quickens the germination of seeds; and assists in the formation of the colouring matter of leaves. Seeds and cuttings, which are required to germinate quickly, will do so under the effect of blue glass (which is equivalent to saying, the effect of an increased proportion of *actinism)*, in half the time they would otherwise require.

95. *In what season of the year is the actinic power of light the greatest?*

In the *spring*, when the germination of plants demands its vitalising aid. In *summer*, when the maturing process advances, *light* and *heat* increase, and *actinism* relatively declines. In the *autumn*, when the ripening period *arrives*, *light* and *actinism* give way to a greater ratio of *heat*.

96. *Why do gardeners tie up the leaves of lettuces, cabbages, &c.?*

Because they find that, by shutting off the rays of light, the leaves that are screened from the chemical effects of the sun's ray's become *etiolated*, or *blanched*. For the same reason they earth up celery, and other plants, the whiteness of which they desire to preserve.

97. *Why is the wood of trees white under the bark?*

Because the bark protects the wood from the effect of the *chemical rays* of light.

98. *Why do potatoes that grow in the light become deteriorated?*

Because the chemical rays of light *produce a change in their substance*, rendering them unfit for food.

99. *Why do the roses of red-rose trees become pale and almost white if deprived of light?*

Because light exercises a chemical agency upon the *colouring of plants* grown under its influence. Sir Humphrey Davy was of opinion that light entered into the composition of plants, and his experiments went to prove that pink, orange, and yellow flowers

"Consider the lilies of the field, how they grow; they toil not, neither do they spin;"

imbibe a smaller portion of light than red ones; that white flowers contain no light; and that vegetables are not only indebted to light for their *colour*, but that *taste* and *odour* are affected by the same source.

100. When we gaze upon a rose it is not its beauty alone that should impress us: every moment of that flower's life is devoted to the fulfilment of its part in the grand scheme of the universe. It decomposes the rays of solar light, and sends the red rays only to our eyes. It absorbs or radiates heat, according to the temperature of the aerial mantle that wraps alike the flower and the man. It distils the gaseous vapours, and restores to man the vital air on which he lives. It takes into its own substance, and incorporates with its own frame, the carbon and the hydrogen of which man has no immediate need. It drinks the dew-drop or the rain-drop, and gives forth its sweet odour as a thanksgiving. And when it dies, it preaches eloquently to Beauty, pointing to the end that is to come.

101. *What is the effect of light on animals?*

Animals in general droop when deprived of light; they *become unhealthy, and die prematurely*. When a man has been long confined in a dark dungeon, his complexion becomes sallow, pustules filled with watery humours break out on his skin, and he becomes dropsical.

102. *Why are worms and grubs that live in the earth, or in the dark, generally of a whitish colour?*

Because, like vegetables deprived of light, they are *blanched* by the absence of the chemical rays of light.

103. *Why are the flowers, fruits, and insects of tropical climates much richer in their colours than those of more temperate regions?*

Because the *greater intensity of the tropical light* impresses with greater power the chemical agency of the sun's rays upon those bodies.

104. *Why are insects that fly by night generally of pale colour?*

Because they are *less affected* by light than those insects that fly by day.

"And yet I say unto you, that even Solomon, in all his glory, was not arrayed like one of these"—MATT. VI.

105. *Why are the backs of fish more or less dark, while the bellies of all of them are white?*

Because their backs are exposed to the light; but their bellies being *always turned from it*, are unaffected by its chemical agency.

106. *Why do persons of fair complexion become freckled in summer time?*

It is supposed that the freckles are produced by the *iron of the blood*, which becomes *fixed* in the skin through the *chemical agency* of the sun's rays.

107. We have had frequently, in the progress of our lessons, to refer to *light* in its connection with the chemistry of nature, and with organic life. Let us now invite the student to pause, and for a moment contemplate the wonders of a sunbeam. How great is its velocity—how vast its power—how varied its parts—yet how ethereal! First, let us contemplate it as a simple beam in which *light* and *heat* are associated. How deep the darkness of the night, and how that darkness clings to the recesses of the earth. But the day beams, and darkness flies before it, until every atom that meets the face of day is lit up with radiance. That which before lay buried in the shade of night is itself now a radiator of the luminous fluid. Mark the genial warmth that comes as the sister of light; then stand by the side of the experimentalist and watch the point on which he directs the shining focus, and in an instant see iron melt and stones run like water, under the fervent heat! Now look upward to the heavens, where the falling drops of rain have formed a natural prism in the rainbow, and shown that the beam of pure whiteness, refracted into various rays, glows with all the tints that adorn the garden of nature. These are the visible effects of light. But follow it into the crust of the earth, where it is, by another power, which is neither light nor heat, quickening the seed into life; watch it as the germ springs up, and the plant puts forth its tender parts, touching them from day to day with deeper dyes, until the floral picture is complete. Follow it unto the sea, where it gives prismatic tints to the *anemone*, and imparts the richest colours to the various *algæ*. Think of the millions of pictures that it paints daily upon the eyes of living things. Contemplate the people of a vast city when, attracted by some floating toy in the air, a million eyes look up to watch its progress. The sun paints a million images of the same object, and each observer has a perfect picture. It makes common to all mankind the beauties of nature, and paints as richly for the peasant as for the king. The Siamese twins were united by a living cord which joined their systems, and gave unity and sympathy to their sensations. In the great flood of light that daily bathes the world, we have a bond of union, giving the like pleasures and inspirations to millions of people at the same instant. And that which floods the world with beauty, should no less be a bond of unity and love.

"Behold how great a matter a little fire kindleth."—JAMES III.

CHAPTER VI.

108. *What is caloric?*

Caloric is an *extremely subtile force*, or property of matter, which resides in all substances, and is the source of what we term heat. *Heat* is *caloric* made apparent, or perceptible to the senses.

109. *Why are the terms caloric and heat made use of with respect to heat?*

The term *caloric* is used to express the *cause* of the physical effects which are attributed to that extremely subtile force, or fluid, which is frequently called heat; but the term *heat* more properly applies to the *sensation*, or *perception*, of heat, as when we say we *feel* the heat of fire, or that a substance is approaching a red heat.

110. *What is latent caloric?*

Latent caloric is that which is *concealed*, or *hidden* in bodies. It may be demonstrated to exist in them, but we have no *ordinary means* of perceiving it. Ice, and water which is as cold as ice, seem, *by the impressions they impart to the touch*, to have the *same temperature*. But there is hidden in the water *a great deal more heat* than there exists in the ice.

111. *What are the principal uses of caloric?*

Caloric is *indispensable* to the existence of animals and vegetables without exception; the entire world of inorganic matter is *affected in all its changes and conditions* by the presence and amount of caloric. It is the element with which man cheers his home, prepares his food, dissolves metals, vitrefies rocks, hardens clay, softens iron, drives the steam engine, and gives to all the productions of the earth the forms and combinations which his necessities require.

112. *Why do we realise the sensation of heat?*

When we put our hand upon a hot body, part of the caloric contained in that body *quits it and enters the hand.* The sensation thus produced we call *heat.*

"For what is your life? It is even a vapour, that appeareth for a little time, and then vanisheth away."—JAMES IV.

113. *Why do we realise the sensation of cold?*

When we touch a cold body, some caloric *leaves our hand, and enters the body we touch.* This causes the sensation which we call *cold.*

114. If you lay your hand upon a *woollen table-cover*, or upon the *sleeve of your coat* or mantle, it will feel *neither warm nor cold*, under ordinary circumstances. But if you raise your hand from the table-cover, or coat, and lay it on the marble mantel-piece, the mantel-piece will feel *cold.* If now you return your hand from the mantel-piece to the table-cover, or coat, *a sensation of warmth will become distinctly perceptible.*

115. *Why have all animals an instinctive dread of fire, although they love warmth?*

Doubtless this instinct has been imparted to them by an all-wise Creator for the purposes of *preservation.* To man *alone* has the use and the control of this powerful and destructive agent been given. If creatures of less intelligence were to meddle with it, *our lives and our possessions would be insecure for a single moment.*

116. *How is caloric produced by combustion or burning?*

In all vapours, gases, and fluids, there exists a large amount of caloric in a *latent*, or *hidden* state. When a substance is burnt, the oxygen gas of the air is decomposed, and the caloric which is contained is set free. Some amount of caloric is also disengaged from whatever solid substance undergoes combustion.

117. *How is caloric produced by percussion?*

Caloric exists between the atoms of all bodies; when, therefore, one body is struck against another, the atoms are compressed, and the caloric may be said to be *squeezed*, or *forced out.* The effect depends upon the nature of the body compressed, the force of compression employed, &c. If the force is great, considerable heat, attended by a spark, will be the result.

118. A blacksmith may hammer a small piece of iron until it becomes *red hot.* With this he might light a match, and *kindle the fire of his forge.* The iron has become more dense by the hammering, and it cannot again be heated to the same degree by similar means, until it has been exposed *in fire*, to *a red heat.* Is it not possible that, by hammering, the particles of iron have been driven closer together, and the *latent heat* driven out? No further hammering will force the atoms nearer, and therefore no further heat can be developed. But when the iron has *again absorbed caloric*, by being plunged in

"Yet man is borne to trouble as the sparks fly upward. I would seek unto God, and unto God would I commit my cause."—JOB V.

fire, it is again charged with latent heat. Indians produce *sparks* by rubbing together *two pieces of wood.* Two pieces of ice may be rubbed together until sufficient warmth is developed to *melt them both.* The axles of railway carriages frequently become *red hot* from *friction.*

119. *Why does evaporation produce cold?*

Evaporation is the *opposite effect* to condensation. When the atoms of bodies are forcibly compressed, the caloric hidden among them is *driven* out; when the atoms of bodies are expanded, as in evaporation, *heat is absorbed.* Hence evaporation is always attended by *cold,* and condensation by *heat.*

120. *Why are wet clothing and damp beds dangerous?*

Because, when the water which causes the dampness evaporates, it takes away a large amount of *caloric from the body,* reducing its temperature below the healthy standard.

121. *Why does eau-de-cologne give relief to persons suffering from feverish headache?*

Because the spirit of the eau-de-cologne is *very volatile;* and when it is spread over the temples, it instantly flies off, bearing away a great amount of *caloric,* and affording a refreshing feeling of coolness.

122. A person might be frozen to death during very warm weather, by pouring upon his body, for some time, sulphuric ether, and keeping him exposed to a thorough draught of air. The winds on the borders of the Persian Gulf are often so scorching that travellers are suddenly suffocated unless they cover their heads with a wet cloth: if this be *too wet,* they immediately feel an intolerable cold which would prove fatal if the moisture was not speedily dissipated by heat.

123. *Why are porous earthenware vessels used for cooling water?*

Because the portion of water which oozes through to their surfaces evaporates therefrom, and bears away the *caloric* of the confined water.

124. *Why, if the "cooler" be placed in the sun, will the water become colder than otherwise?*

Because the heat of the sun, instead of passing through the cooler, is carried away by the rapid evaporation of water from the surface of the vessel; and in this *rapid evaporation*, the caloric of the confined water is also partly drawn off.

The blacks in Senegambia have a similar method of cooling water by filling tanned leather bags with it, which they hang up in the sun; the water oozes through the leather so as to keep the outward surface wet, which, by its quick and continued evaporation, cools the water remarkably.

125. *Why does the collision of flint and steel produce sparks?*

The *density* of the substances of the flint and steel facilitates the driving out of caloric by a sharp blow. If the particles of those bodies were soft and yielding, the caloric would simply be *driven from one part of their substance to another*. But as their atoms cannot easily be disturbed, the compression caused by a smart blow drives the caloric out at the surface, as the *readiest point of escape*.

126. *Why must the blow be very instantaneous to produce a spark?*

Because the substances of both flint and steel, and especially of the steel, are *good conductors*, and would quickly conduct away the excess of heat produced at any one point. But the smartness of the blow, while it compresses the atoms of those bodies in the highest degree, produces heat so suddenly that it cannot escape by conduction.

127. What an eloquent lecture might be delivered upon the old-fashioned Tinder-box, illustrated by the one experiment of "striking a light." In that box lie, cold and motionless, the Flint and Steel, rude in form and crude in substance. And yet, within the breast of each, there lies a spark of that grand element which influences every atom of the universe; a spark which could invoke the fierce agents of destruction to wrap their blasting flames around a stately forest, or a crowed city, and sweep it from the surface of the world, or which might kindle the genial blaze upon the homely hearth, and shed a radiant glow upon a group of smiling faces; a spark such as that which rises with the curling smoke from the village blacksmith's forge—or that which leaps with terrific wrath from the troubled breast of a Vesuvius. And then the Tinder—the Cotton—the Carbon; what a tale might be told of the cotton-field where it grew, of the black slave who plucked it, of the white toiler who spun it into a garment, and of the village beauty who wore it—until, faded and despised, it was cast among a heap of old rags, and finally found its way to the tinder-box. Then the Tinder might tell of its hopes, how, though now a blackened mass, soiling everything that touched it, it would soon be wedded to one of the great ministers of nature, and fly away on transparent wings, until, resting upon some Alpine tree, it would make its

"Yea, I will gather you, and blow upon you in the fire of my wrath, and ye shall be melted in the midst thereof."—EZEKIEL XXIII.

home among the green leaves, and for a while live in freshness and beauty, looking down upon the peaceful vale. Then the Steel might tell its story, how for centuries it lay in the deep caverns of the earth, until man, with his unquiet spirit, dug down to the dark depths and dragged it forth, saying, "No longer be at peace." Then would come tales of the fiery furnace, what Fire had done for Steel, and what Steel had done for Fire. And then the Flint might tell of the time when the weather-bound mariners, lighting their fires upon the Syrian shore, melted silicious stones into gems of glass, and thus led the way to the discovery of the transparent pane that gives a crystal inlet to the light of our homes; of the mirror in whose face the lady contemplates her charms; of the microscope and the telescope by which the invisible are brought to sight, and the distant drawn near; of the prism by which Newton analysed the rays of light; and of the photographic camera in which the sun prints with his own rays the pictures of his own adorning. And then both Flint and Steel might relate their adventures in the battle-field, whither they had gone together, and of fights they had seen in which man struck down his fellow-man, and like a fiend had revelled in his brother's blood. Thus, even from the cold hearts of Flint and Steel, man might learn a lesson which should make him blush at the "glory of war;" and the proud, who dispise the teachings of small things, might learn to appreciate the truths that are linked to the story of a "tinder-box."

128. *Why is caloric produced by friction?*

Because, as in the case of the flint and steel, the compression of the particles of the bodies between which the friction is produced, *sets free their latent caloric.*

But, in the case of friction, another fact must be considered, which is, that caloric has a tendency *to flow towards a point of excitement.* Thus the first heat produced by friction would arise from the escape of the caloric of the surfaces rubbed together; but as the action continued, the heat would increase, because caloric would begin to flow from the more distant portions of the excited bodies to the point of excitement.

129. *Why do forests sometimes take fire?*

Because in dense woods the branches of trees, when moved by the violence of the winds, are rubbed together, producing the degree of *friction* necessary to ignite them.

130. *How can caloric be disengaged by electricity or galvanism?*

A more intense degree of caloric may be obtained by the aid of an electrical battery, or by the galvanic apparatus, than by any other means. But the *relation between heat and electricity is not at present understood.* Electricity produces heat, and heat electricity—the "reason why" is one of the grand problems remaining open for solution by those master-minds that love to pursue the theorems of scientific inquiry.

131. *Why is caloric produced by chemical mixture?*

Because when chemical action takes place, the atomic condition of bodies undergoes a change. Whenever fluid particles assume a more solid form, *heat is given off.* When water is poured upon quick-lime, the water unites to the lime and forms a solid body; the caloric *which caused the fluidity of the water* is given off, and a great amount of heat is evolved.

132. *What is the caloric which exists in bodies usually called?*

It is called *latent caloric*, or, sometimes, the *caloric of fluidity.*

133. *Why is it sometimes called the caloric of fluidity?*

Because, wherever it exists *largely*, the body in which it exists is in a *fluid condition*, and this condition is attributed to the presence of the amount of caloric.

134. *Why is caloric sometimes called "free?"*

The term *free caloric* is applied to it whenever it is uncombined with any substance. If we heat a poker until it becomes red hot, that amount of caloric which is taken up by the poker, in excess of what was latent in it before, may be regarded as *free caloric;* for, as it is not combined with the iron, it will quit it directly that the poker is removed from the fire, and will diffuse itself by passing from the poker by radiation, or by other means, and entering into whatever colder bodies may come within the reach of its influence.

135. *How can latent caloric and free caloric exist in a body at the same time?*

This may be easily explained by the familiar example of a piece of common bread which has been dipped in water. The bread will contain two portions of water—one of them in a state of *combination*, forming a part of the bread; the other only *interposed between the particles* of the bread, and capable of being squeezed out by pressure.

"As smoke is driven away, so drive them away; as wax melteth before the fire, so let the wicked perish at the presence of God."—PSALM XVIII.

136. *What substances contain latent caloric?*

Caloric in a latent state exists in all *substances in nature.*

137. *Do all substances contain the same amount of latent caloric?*

No; caloric combines with different substances in very different proportions; and for this reason, one body is said to have *a greater capacity* for caloric than another.

138. *What is meant by the term "capacity for caloric?"*

If a lock of wool and a piece of sponge of equal size be dipped in water, the sponge will *take up a larger amount of water than the wool,* and hence it may be said that the sponge has *a greater capacity for water* than the wool.

139. *Is the capacity for caloric uniformly the same in the same bodies?*

The same bodies have at all times *an equal capacity for caloric;* but when a *change* occurs in the state of those bodies, then their capacity alters.

Whenever a body changes its chemical state, it either combines with, or separates from, caloric.

140. *Why, in making butter, does it become warm at the moment when the cream changes from a fluid to a solid?*

Because the heat, which was latent in the cream in its fluid state, becomes in some measure *in excess and free* when it passes into the solid state; it is therefore given off, and made perceptible.

141. *What are the effects of caloric upon bodies?*

The general effects of caloric are to *increase the bulk* of the substances with which it unites, and to render them specifically lighter than they were before; but in whatever quantity it is accumulated in bodies, it never adds to their absolute weight.

142. *How does caloric act upon hard bodies to convert them into fluids?*

It *insinuates itself among their particles,* and separates them in some measure from each other. Thus, ice is converted into water, and, by a further addition of caloric, into steam.

"The fining pot is for silver, and the furnace for gold: but the Lord trieth the hearts."—PROVERBS XVII.

143. *Why are clocks and watches frequently affected by a change of temperature?*

Because the metals of which their machinery is composed are expanded by heat, and contracted by cold, causing them to *vary* in their indications of time.

144. *Why do stringed instruments become out of tune when untouched for some time?*

Because changes of temperature, acting upon the several parts, some of which are composed of wood, and others of metal, which expand in *different degrees*, interfere with the exactness of adjustment necessary to produce musical sounds.

145. *Why do glasses crack if suddenly heated?*

Because the heat, expanding their surfaces suddenly, before it can reach the other parts, produces *an unequal tension of the substance*, which therefore splits.

146. *Why are cut glasses more liable to crack with sudden heat than plain ones?*

Because of the *unequal thickness* of their parts; the thin parts expand more rapidly than the thick ones.

147. *Why are the retorts used by chemists in their experiments made of thin glass?*

Because the thinness of their substance favours the rapid and equal diffusion of heat; they are therefore less liable to crack than if they were thicker.

148. *Why does the heater of an Italian iron frequently become too large for its case?*

Because the atoms of the iron are held apart by the repulsive agency of the caloric.

149. *Why has caloric been called a repulsive agent?*

Because it *repels the atoms* of which bodies are composed, and tends to counteract the force of cohesive attraction, by which they are held together in solid compact.

"Their horses' hoofs shall be counted like flint, and their wheels like a whirlwind."—ISAIAH V.

150. Some years since the walls of the *Conservatoire des Arts et Metiers, at Paris*, were found to be diverging from the perpendicular. They were restored to their original lines by the following beautiful expedient:—They were perforated transversely, copper bars were thrust through the perforations, each bar at either extremity being supplied with a nut and screw. Every alternate bar was now heated by means of a spirit-lamp flame; being heated the bars expanded, and the screw-nuts being now turned close up to the walls on either side, the bars were allowed to cool. By cooling they contracted, pulled the walls to some extent together, leaving the ends of the unheated bars protruding; their screw-nuts were now turned close up to the wall on either side, and the heating process repeated. Thus little by little the walls were restored to their original position.

This effect is frequently observed by females in domestic life, who, when they are ironing, or using the Italian irons, find that the heated metal has been too much expanded to enter the box or tube. They find it necessary to wait until the cooling of the iron has had the effect of reducing its dimensions. The expansion of bodies by heat is one of the grandest and most important laws of nature. We are indebted to it for some of the most beautiful, as well as the most awful phenomena. And science has gained some of its mightiest conquests through its aid. Yet frequently, though quite unthought of, in the hands of the humble laundress, will be found a most striking illustration of this wonderful force of caloric.

CHAPTER VII.

151. *What is the chief source of caloric?*

The *sun* is the chief, and, probably, the original fountain which furnishes the earth with a regular supply, and renders it capable of supporting the animal and vegetable creations.

152. The amount of heat which our *earth receives from the sun*, and the economy of that heat by the laws of *radiation*, *absorption*, *convection*, and *reflection*, are exactly proportionate to the necessities of our planet, and the living things that inhabit it. It is held by philosophers that any change in the orbit of our earth, which would either increase or decrease the amount of heat falling upon it, would, of necessity, be followed by the *annihilation of all the existing races*. The planets Mercury and Venus, which are distant respectively 37 millions of miles, and 68 millions of miles, from the great source of solar heat, possess a temperature which would *melt our solid rocks;* while Uranus (1,800 millions of miles) and Neptune (whose distance from the sun has not been determined) must receive so small an amount of heat, that water, such as ours, would become as solid as the hardest rock, and our atmosphere would be resolved into a liquid! Yet, poised in the mysterious balance of opposing forces, our orb flies unerringly on its course, at the rate of 65,000 miles an hour; preserving, in its wonderful flight, that precise relation to the sun, which causes it to receive from his life-inspiring rays the exact degree of heat, which, being shared by every atom of matter, and every form of organic existence, *is just the amount needed to constitute the heat-life of the world!*

"I will put my laws into their hearts, and in their minds I will write them."—
HEBREWS X.

153. *Why do the axles of coach wheels become hot when not sufficiently greased?*

Because the friction of the surfaces upon each other creates a disturbance of the surrounding particles, and *sets free the caloric* which lies latent among them.

154. *Why do the axles of the front wheels become hotter than those of the 'hind ones?*

Because, being much smaller than the others, they *revolve more rapidly;* therefore a greater amount of friction takes place upon their axles.

155. *Why does the application of grease or oil prevent the development of heat in surfaces that undergo friction?*

Because the superior mobility of the particles of fat or oil enables them *to move freely over each other,* without the atomic disturbance which results when dry and hard surfaces move upon each other.

156. *What are the laws that govern the diffusion of heat?*

They are—1. Conduction; 2. Radiation; 3. Convection; 4. Reflection.

157. *What is the conduction of heat?*

The *conduction of heat* is its transmission through a substance, the particles of which are fixed, either by penetrating those particles and being transmitted from one to another, or by permeating the inconceivably small pores that exist even in the densest bodies.

Water has been *forced through gold* by pressure. We may, therefore, comprehend how so subtile an essence as caloric may permeate the most solid bodies.

158. *What is the radiation of heat?*

The *radiation of heat* is its diffusion in rays to bodies surrounding it, precisely as light is thrown off in all directions from a luminous body.

159. *Is radiated heat transmitted by the surrounding atmosphere?*

No; heat is as readily radiated from a hot body in a *vacuum* as in air. The radiation of heat and light is a peculiar and special law, independent of other laws that govern the dispersion of the elements.

"Drought and heat consume the snow waters; so doth the grave those which have sinned."—JOB XXIV.

160. *What is the convection of heat?*

The *convection of heat* is its distribution by the movement of the particles of air or water, which, becoming heated by contact with the calorific body, rise by virtue of their expansion, and give place to colder particles that in their turn perform the same office. It is, in fact, the *conveyance* of heat by moving atoms.

161. *Why does a soap-bubble ascend when first blown?*

Because, being filled with *warm air* from the lungs, it is *lighter* than the surrounding air upon which it rises.

162. *Why does it descend after a time?*

Because, having *parted with its warmth* in its passage through the air, it has become relatively heavier, and therefore falls to the ground.

This *conveyance of heat* through the atmosphere by the soap-bubble clearly illustrates the important law of *convection*. Every atom of warm air, or of warm water, as it rises with a glow of heat in its embrace, and diffuses that heat to surrounding atoms, performs the part which the soap-bubble has made familiar to our comprehension.

163. *What is the reflection of heat?*

The *reflection of heat* is the turning back of its rays by surfaces upon which they fall; and is in many respects similar to the reflection of light.

164 There is a curious and an exceptive fact with reference to *reflected* heat, for which we confess that we are unable to give "*the reason why*." It is found that snow, which lies near the trunks of trees, or the base of upright stones, melts before that which is at a distance from them, though the sun may shine equally upon both. If a blackened card is placed upon ice or snow under the sun's rays, the frozen body underneath it will be thawed before that which surrounds it. But if we *reflect* the sun's rays from a metal surface, the result is *directly contrary*—the exposed snow is the first to melt, leaving the card standing as upon a pyramid. Snow *melts* under heat which is *reflected* from trees or stones, while it withstands the effect of the *direct solar rays*. In passing through a cemetery this winter (1857), when the snow lay deep, we were struck with the circumstance that the snow in front of the head-stones facing the sun was completely dissolved, and, in nearly every instance, the space on which the snow had melted assumed a coffin-like shape. This forced itself so much upon our attention that we remained some time to endeavour to analyse the phenomenon, and it was not until we remembered the curious effect of

" What time they wax warm, they vanish when it is hot, they are consumed out of their place."—JOB VI.

reflected heat that we could account for it. It is obvious that the rays falling from the upper part of the head-stone on to the *foot* of the grave would be less powerful than those that radiated from the *centre* of the stone to the centre of the grave. Hence it was that the heat dissolved at the foot of the grave only a narrow piece of snow, which widened towards the centre, and narrowed again as it approached the foot of the head-stone, where the lines of radiation would naturally decrease. Such a phenomenon would prove sufficient to raise superstition in untutored minds.

165. *Why are some bodies called conductors of heat?*

Because they are found to *afford a passage to caloric.*

166. *Why are other substances called non-conductors of heat?*

Because they *less readily permit the passage of caloric,* and in some cases they almost prevent it.

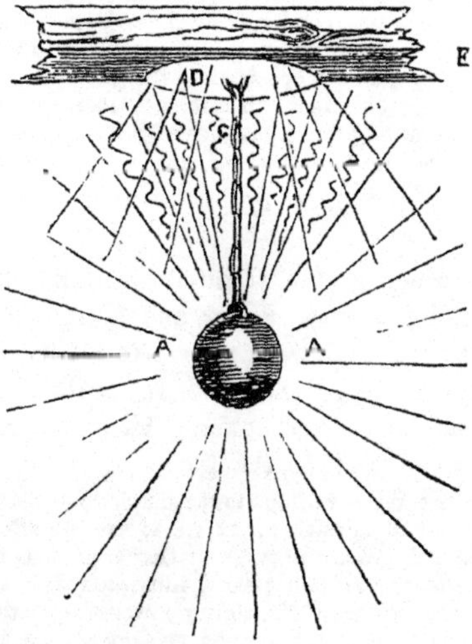

DIAGRAM ILLUSTRATING THE DISTRIBUTION OF HEAT.

167. The accompanying engraving will be found to illustrate *all the laws which govern the distribution of heat.* It represents a cannon ball, supposed to be red-hot. The *straight lines,* A, that start out in every direction from the ball, represent the lines of *Radiant* heat; the *waved lines* represent the tremulous motion imparted to the air as it rises from the ball, heated by contact with its surface, and diffuses that heat by *Convection;* the links of the chain, C, become

"And I heard as it were the voice of a great multitude, and as the voice of many waters, and as the voice of mighty thunderings, saying alleluia: for the Lord God omnipotent reigneth."—REV. XIX.

gradually hot by the transmission of caloric, through their substance, or by *Conduction;* D represents a metal screen *reflector*, which has the effect of *reflecting* back the radiant heat; the staple is fastened into a wooden beam, E, which, being a *non-conductor*, arrests the *escape of the heat by conduction* further than the *staple* which holds the chain.

168. *What is the absorption of heat?*

When heat is absorbed, it is *taken up and retained*, so long as the surrounding temperature is equal to that of the absorbing body. But when the external temperature falls, the body which had absorbed the heat *throws it off by radiation.* It has been found that the *colours* of bodies affect their powers of absorption and radiation.

169. If, when the ground is covered with snow, pieces of woollen cloth, of equal size and thickness, and differing only in colour, are laid upon the surface of the snow, near to each other, it will be found that the relation of *colour* to temperature will be as follows:—In a few hours the *black* cloth will have dissolved so much of the snow beneath it as to sink deep below the surface; the *blue* will have proved nearly as warm as the black; the *brown* will have dissolved less of the snow; the *red* less than the brown; and the *white* the least, or none at all. Similar experiments may be tried with reference to the *condensation of dew*, &c. And it will be uniformly found that the *colour* of a body materially affects its powers of *absorption* and of *radiation*.

170. *Why do draughts of air sweep across apartments towards the fire?*

Because, no sooner is fuel lighted in a fire-place, than the superincumbent air becomes *hotter* and *specifically lighter;* it, ascends, therefore, and cold air rushes in to fill its place. Thus we have in point of fact a local wind; and we see illustrated by our own fire-side the great law of *convection* which exercises so great an influence over every part of the globe.

171. To the educated mind, nothing is too simple to merit attention. To the ignorant, few things are sufficiently attractive to excite curiosity. Knowledge enables us to estimate the varied phenomena that are hourly arising around us, and to see, even in the most trifling effects, illustrations of those great causes and consequences that govern with mighty power the material world. Man, sitting by his fire-side, is enabled to witness the operation of some of nature's grandest laws: *light* and *heat* are around him; *conduction, radiation, reflection, absorption,* and *convection* of heat are all going on before him; little winds are sweeping by his footstool, and warm currents, with miniature clouds folded in their arms, are passing upward before his view. Chemical changes are going on; the solid rock of coal disappears, flying away as an invisible gas. The little "hills are melted," and hard stones have been converted into "fervent heat." Although some of these changes are imperceptible to the *eye*, they are manifest to the educated *mind*, and the pleasures of philosophical observation are as sweet as a poet's dreams.

"Lo, these are parts of his ways: but how little a portion is heard of him? but the thunder of his power who can understand."—JOB XXVI.

CHAPTER VIII.

172. *Why does the rubbing of a stick of sealing-wax cause it to attract small particles of matter?*

Because it excites in the sealing-wax that *force* which was first observed in the *amber*. Sealing-wax, therefore, is called an *electric (amber-like)* body.

173. *Why is electricity termed the electric fluid?*

Simply because the term *fluid* is the most convenient that can be found to express our ideas when speaking of the *phenomena of electric force*. But of the nature of electricity, except through its observed *effects*, nothing is known.

174. *What substances are electric?*

All substances in nature, from the *metals* to the *gases*. But they differ very widely in their electrical qualities.

175. *What is positive electricity?*

Electricity, when it exists, or is excited, in any body, to an amount which is *in excess* of the amount natural to that body, is called *positive* (called also *vitreous*).

176. *What is negative electricity?*

Electricity, when it exists, or is excited, in any body, in an amount which *is less* than is the amount natural to that body, is called *negative* (called also *resinous*).

177. *Why is "positive" electricity called also "vitreous," and "negative" electricity called also "resinous?"*

Because some philosophers believe that there is but *one electricity*, but that it is liable to variations of *quantity* or *state*, which they distinguish by *positive* and *negative;* while others believe that there are *two electricities*, which they name *vitreous* and *resinous*, because they may be induced respectively from *vitreous* and *resinous* substances, and are found display forces of attraction and repulsion.

"The clouds poured out water; the skies sent out a sound, thine arrows also went abroad."

178. *Upon what do the electrical phenomena of nature depend?*

Upon the tendency of *electricity* to find an *equilibrium* between its *positive* and *negative* states (assuming there to be but *one* fluid); or upon the tendency of *vitreous electricity* to seek out and combine with *resinous electricity* (assuming that there are *two* fluids).

179. *How does the equilibrium of electricity become disturbed?*

By changes in the condition of matter. As electricity resides in all substances, and is, perhaps, an essential ingredient in their condition, so every change in the state of matter—whether from heat to cold, or from cold to heat; from a state of rest to that of motion; from the solid to the liquid, or the aeriform condition, or *vice versa;* or whether substances combine chemically and produce new compounds—in every change *the electrical equilibrium is disturbed;* and, in proportion to the degree of disturbance, is the force exerted by electricity to resume its balance in the scale of nature.

180. *How does electricity seek to regain equilibrium?*

By passing through substances that are favourable to its diffusion; therefore they are called *conducting* or *non-conducting* bodies, according as they favour or oppose the transmission of the electrical current.

181. *What substances are conductors of electricity?*

Metals, charcoal, animal fluids, water, vegetable bodies, animal bodies, flame, smoke, vapour, &c.

182. *What substances are non-conductors?*

Rust, oils, phosphorous, lime, chalk, caoutchouc, gutta percha, camphor, marble, porcelain, dry gases and air, feathers, hair, wool, silk, glass, transparent stones, vitrefactions, wax, amber, &c. These bodies are also called *insulators*. Some of these substances, as chalk, feathers, hair, wool, silk, &c., though non-conductors when *dry*, become conductors when *wetted*.

Insulating - preventing from escape.

"The voice of the Lord is upon the waters: the God of Glory thundereth; the Lord is upon many waters."—PSALM XXIX.

183. *Why are amber and wax classed among the non-conductors, when they have been pointed out as electrics, and used to illustrate electrical force?*

It is because they are *non-conductors* that they display under excitement, the attractive force shown in respect to the particles of matter which were drawn towards their substances. If a bar of *iron* were excited, instead of a stick of wax, electricity would be equally developed; but the iron, *being a good conductor*, would pass the electricity to the hand of the operator as fast as it accumulated, and the equilibrium would be undisturbed.

184. *What is the effect when electricity, in considerable force, seeks its equilibrium, but meets with insulating bodies?*

The result is a violent action, in which *intense heat and light* are developed, and in the evolution of which *the electric force becomes expended.*

185. *What is the cause of electric sparks?*

The electric force, passing through a conducting body to find its *equilibrium*, is checked in its course by an insulator, and emits a spark.

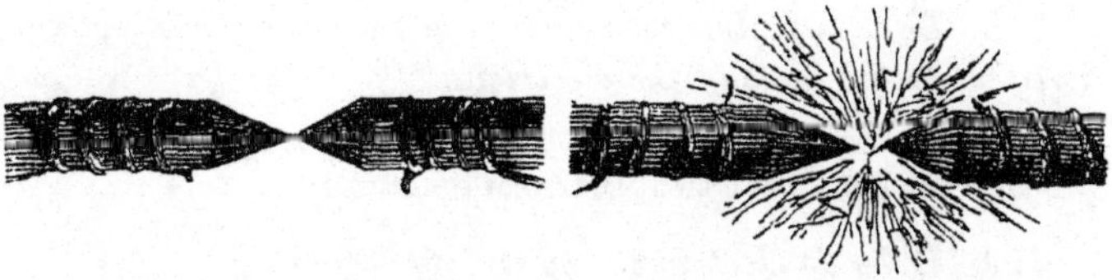

SHOWING THE EFFECT OF THE UNION AND THE SEPARATION OF THE CHARCOAL POINTS.

186. *What produces the electric light?*

Currents of electricity pass towards each other along wires at the ends of which two charcoal points are placed. As long as the charcoal points remain in contact, the electric communication is complete, and no light is emitted, but when they are drawn apart, intense heat and light are evolved.

187. *What is the cause of lightning?*

Lightning is the result of *electrical discharges* from the *clouds.*

"O Lord, our Lord, how excellent is thy name in all the earth! who hast set thy glory above the heavens."—PSALM VIII.

188. *What developes electricity in the clouds?*

Evaporations from the surface of the earth; changes of temperature in the atmospheric vapour; chemical action on the earth's surface; and the friction of volumes of air of different densities against each other.

189. *Why do these phenomena produce electricity?*

Because they disturb the equilibrium of the electric force, and produce *positive* and *negative* states of electricity.

190. *When does lightning occur?*

When clouds charged with the *opposite electricities* approach. the forces rush to each other, and combine in a state of equilibrium.

191. *Why does lightning attend this movement of the forces of electricity?*

Because the atmosphere, being unable to convey the great charges of electricity as they rush towards each other, *acts as an insulator*, and lightning is caused by the *violence of the electricity in forcing its passage.*

192. *Does lightning ever occur when the conducting power is equal to the force of the electricity?*

No; electricity passes invisibly, noiselessly, and harmlessly. whenever it finds a sufficient source of *conduction.*

193. *Why do electric storms purify the air?*

Because they restore the *equilibrium of electricity* which is essential to the salubrity of the atmosphere; they intermix the *gases of the atmosphere*, by agitation; they *precipitate the vapours* of the atmosphere, and with the precipitation of vapours, *noxious exhalations* are taken to the earth, where they become absorbed· they also contribute largely to the formation of *ozone*, which imparts to the air corrective and restorative properties.

"Ye fearful souls fresh courage take,
The clouds ye so much dread
Are big with mercy, and will break
With blessings o'er your head!"

"The voice of thy thunder was in the heaven: the lightnings lightened the world; the earth trembled and shook."—PSALM LXVII.

194. *What is ozone?*

Ozone is an *atmospheric element* recently discovered, and respecting which differences of opinion prevail. It is generally supposed to be *oxygen* in a state of *great strength*, constituting a variety of form or condition.

195. *Why do we know that electricity contributes to the formation of ozone?*

Because careful observations have established the fact that the proportion of *ozone* in the atmosphere is *relative to the amount of electricity*.

196. *What are the properties of ozone?*

It displays an extraordinary power in the neutralisation of putrefactions, rapidly and thoroughly counteracting noxious exhalations; it is the most powerful of all *disinfectants*.

197. Schonbien, the discoverer of *ozone*, inclines to the opinion that it is a *new chemical element*. Whatever it may be, there can be no doubt that it plays an important part in the economy of nature. Its *absence* has been marked by pestilential ravages, as in the *cholera* visitations; and to its *excess* are attributed epidemics, such as *influenza*. It was found, during the last visitation of cholera, that the *fumigation of houses with sulphur* had a remarkable efficacy in preventing the spread of the contagion. The combustion of sulphur ozonised the atmosphere; the same result occurs through the emission of *phosphoric vapours*; ozone is also developed by the electricity evolved by the *electrical machine*, and in the greater *electrical phenomena of nature*. The smell imparted to the air during an electric storm is identical with that which occurs in the vicinity of an electrical apparatus—it is a *fresh* and *sulphurous* odour. The opinion is gaining ground that the respiration of animals and the combustion of matter are sources of ozone, and that plants produce it when under the influence of the direct rays of the sun. It is also believed to be produced by water, when the sun's rays fall upon it. The most recent opinion respecting ozone is, that it is *electrized oxygen*. The subject is of vast importance, and opens another field of discovery to the pioneers of scientific truth.

198. *Why does electricity accumulate in the clouds?*

Because the clouds are *conductors*, but the air surrounding them is a *non-conductor*; when, therefore, electricity is excited in the atmosphere by any natural cause, it is received by the clouds; it is probably this electric charge which prevents the water vescicles from uniting together and falling down in the form of rain.

"He made darkness his secret place: his pavilion round about him were dark waters and thick clouds of the skies."

199. *Why do different clouds become charged with the opposite electricities?*

When two bodies are rubbed together they become electrified—one of them *positively*, and the other *negatively*. It is very probable that when two currents of dry air move in different ways, the friction of the two surfaces may evolve electricity. Clouds floating in the locality of the excitement would receive the electricity, and thus one cloud may become charged with *positive*, and the other with *negative*, electricity.

200. *Why do clouds, when electrified, move towards each other?*

Because bodies which are charged with the opposite electricities *attract each other*—the electricities always seek to establish an equilibrium, and hence two electrified clouds would attract each other.

TWO CLOUDS, CHARGED WITH THE OPPOSITE ELECTRICITIES—(A) POSITIVE AND (B) NEGATIVE.

201. Let it be assumed that the cloud A becomes positively electrified—that is to say, charged with positive electricity. There is not in all nature, and there *cannot be*, such a condition as that of one body *positively* excited without the co-existence of another body *negatively* excited. Hence, if cloud B were away, and cloud A positively excited, the air circumjacent to A would assume the second or negative function; but if the cloud B is present, it therefore becomes negative, and the two clouds A and B are mutually attracted, because opposite electricities, attract each other. Hence they approach until the space of air between the two is insufficient to restrain their mutual electric tension: this condition having arrived, a discharge takes place.

"The Lord also thundered in the heavens, and the Highest gave his voice; hailstones and coals of fire."

202. *Why does a flash of lightning occur when the electrified clouds approach each other?*

Because the air between the clouds is a non-conductor; it is the force of electricity *overcoming the resistance of the atmosphere* which occasions the *flash of lightning.*

203. *Why does a shower of rain generally succeed lightning?*

Because the equilibrium of a certain amount of electricity having been restored, the clouds, deprived of their electricity, *collapse into rain.*

204. *Why does a thunderstorm sometimes cease after a few flashes, and a smart shower?*

Because when the electrical exchanges occur only between clouds, the *equilibrium* of their electricities is *soon restored.*

205. *Why does a thunderstorm at other times continue for a long period?*

Because *the air* as well as the *clouds*, is involved in the electrical disturbance. The air with which a cloud comes in contact, being a non-conductor, would not lose its electricity by the discharge of the cloud, but would *continue to supply the cloud with new charges;* and this repeated charging and discharging would continue till the different strata of excited air were brought to their natural state.

206. *Does lightning ever pass from the air to the earth, and from the earth to the air?*

Thunderstorms usually take place between the clouds, or different strata of air. But sometimes when clouds charged with an opposite electricity to that of the earth, or of a mountain, approach it, a discharge takes place *from the cloud to the earth, or from the earth to the cloud.*

207. The mingling of the electricities of the earth and the air must be continually going on. But *lightning* does not attend the phenomenon, because all natural bodies, vapours, trees, animals, mountains, houses, rocks, &c., act more or less as conductors between the earth and the air. It is only when there is a great disturbance of the electrical forces, that *terrestrial lightning* is developed. When lightning strikes the earth with great force, it sometimes produces what are called *fulgurites* in sandy soils; these are hollow tubes, produced by the melting of the soil.

"Yea, he sent out his arrows, and scattered them; and he shot out lightnings and discomfited them."—PSALM XVIII.

208. *Why does the peal of thunder occur after the flash of lightning?*

The flash and the thunder are really simultaneous; bu as light travels with a velocity immensely greater than that of scund, *we see the flash* some time before we *hear the thunder.*

209. *How may we calculate the distance at which the electric discharge takes place?*

Sound travels at the rate of *a quarter of a mile in a second.* If, therefore, the peal of thunder is heard *four seconds* after the flash of lightning, the discharge took place about a mile off. The pulse of an adult person beats about *once in a second;* therefore, guided by the pulse, any person may calculate the probable distance of the storm:—

2 beats, ½ a mile.
3 beats, ¾ of a mile.
4 beats, 1 mile.
5 beats, 1¼ mile.
6 beats, 1½ miles.
7 beats, 1¾ miles.
8 beats, 2 miles, &c.

Attention should be paid to the *direction and speed of the wind,* and some modifications of the calculation be made accordingly. Persons between 20 and 40 years of age should count *five beats of the pulse to a mile;* under 20, *six beats.*

210. *What is the extent of mechanical force of lightning?*

Lightning has been proved to have struck a church, St. George's Church, Leicester, on the 1st of August, 1846, with a force equal to more than 12,000 horse-power. A single horse-power, in mechanical calculations, is equivalent to raising a weight of 32,000 lbs. one foot in a minute. The force of lightning, therefore, has been proved to be equal to the raising of 384,000,000 lbs. one foot in a minute. This is equal to the united power of twelve of our largest steamers, having collectively 24 engines of 500 horse-power each. The velocity of electricity is so great that it would travel round the world *eight times in a minute.*

"Can any understand the spreadings of the clouds, or the noise of his tabernacle?"—JOB XXXVI.

211. *What gives the varying character to the flashes of lightning?*

Lightning is *zig-zag* when it travels through a long distance, because it *compresses the air*, which interferes with its direct course.

It is *straight* when it passes through a short distance only.

It is *forked* when, being resisted by the air, it divides into two or more points.

It is *sheet* when the flash is distant, and is seen by reflection in distant parts.

It is *blue* when the electrical excitement is very intense.

212. *What is thunder?*

Thunder is the *noise which succeeds the rush* of the electrical fluid through the air.

213. *Why does noise follow the commotion caused by electricity?*

Because, by the violence of the electric force, vast *fields of air are divided*; great volumes of air are *rarefied*; and *vapours* are *condensed*, and thrown down as *rain*. Thunder is therefore caused by the *vibrations of the air* as it collapses and seeks to restore its own *equilibrium*.

214. *What gives the varying character to the sounds of thunder?*

Its peals are most tremendous in mountainous regions. When interrupted in their advance by hills, or other elevated objects, the reverberation of the peals is broken and irregular.

They consist of a single and sudden clap when *the storm is near*, and when the *country is level*.

They are rattling and rumbling when the forked lightning occurs *in different directions and distances*.

215. *Why is lightning sometimes unattended by thunder?*

The absence of thunder sometimes arises from the *great distance* of the storm; at other times from the *nearness of the clouds to each other* at the moment of the discharge, occasioning but a slight disturbance of the atmosphere.

"His lightnings enlightened the world: the earth saw and trembled."—
PSALM XCVII.

CHAPTER IX.

216. *Why is it necessary to avoid proximity to water and streams during lightning?*

Because water being a good conductor, a person standing near it, being also a good conductor, might determine the discharge.

217. *Why is the middle story of a house the safest part to be in during an electric storm?*

Because lightning sometimes *passes upward from the earth;* hence it is a mistaken notion of safety to take refuge in a cellar. In instances where houses have been struck with electricity from the earth, the force of the shock has abated as the electricity ascended.

218. *Why is the middle of a room the safest place?*

Because it is removed from *conducting surfaces,* such as bright mirrors, bell-wires, walls, cornices, &c.

219. *Why is it advisable to stand on a thick hearth-rug, or on a hair mattress?*

Because, being dry, and non-conducting bodies, they would *insulate the human body,* and prevent the electricity from passing to it.

220. *Why is it dangerous to approach the fire-place in an electric storm?*

Because heated air, smoke, vapour, and soot are conductors; and the metal grate, fender, fire-irons, &c., are all good conductors, and *may conduct the electricity to a person standing near them.*

221. *Why is it dangerous to approach a window for the purpose of fastening the shutters, during a thunderstorm?*

Because the iron bolts and bars, the hinges, &c., *afford conduction to electricity,* and may convey a severe electric shock.

222. *Why should a person riding in a carriage during a thunderstorm, sit upright?*

Because electricity may pass through the sides or back of the carriage.

"After it a voice roareth: he thundereth with the voice of his excellency; and he will not stay them when his voice is heard."—JOB XXXVII.

223. The rule should be observed to avoid all *elevated* objects, which, as well as being elevated, are of a *slow conducting* character. The advice has been frequently given to avoid the neighbourhood of conducting bodies; but when those bodies offer *perfect conduction* they convey electricity to the ground. Thus, to stand by or be within a church or lofty building, *properly fitted with lightning conductors*, is far safer than to stand near a tree, or a lofty building, *without a lightning conductor*, because electricity always seeks *the best conductor*, and the human body being a better medium than either a tree or a house, electricity would fly to it, even though it may first have seized upon one of the other objects. It must not be supposed that churches and houses having lightning conductors are not touched by electricity because no shock is felt or witnessed: their lightning rods are continually facilitating the movements of electricity, and probably passing, harmlessly, currents which, if resisted, would shatter the edifices to pieces.

224. *Why are lightning conductors attached to tall edifices, monuments, &c.?*

Because they convey the electricity of the air and clouds *harmlessly* to the earth.

MONUMENT AND LIGHTNING CONDUCTOR.

225. *Why does copper form the best lightning conductor?*

Because it has been found that electricity passes over a copper surface *more rapidly than over any other.*

226. *Why are lightning conductors made to project above the objects they are intended to protect?*

Because it was thought that by so doing they would *meet* the electricity, and *attract it from the object.* But this is found to be an error; a conductor having its commencement simply on a level with the highest part of a building would be just as effective as if it extended into the air. It is, therefore, unnecessary to disfigure buildings and monuments by having long metal rods projecting from them.

"He directeth it under the whole heavens, and his lightning unto the ends of the earth."

227 *Why do they convey electricity harmlessly to the earth?*

Because electricity flies along their substance with great velocity; therefore it passes harmlessly away. It is only when its progress is *resisted* by a *non-conducting body* that electricity manifests *force.*

228. *Do lightning conductors attract electricity?*

They do not *attract electricity,* they merely afford it a *rapid conduction* when it happens to reach them. They no more attract electricity than water-pipes attached to a house attract water.

229. *Why should a large building have several conductors?*

Because electricity may strike upon *any part* of a building. It is, therefore, proper to have several conductors, and to ramify their branches over the surface of the edifice, so as to form *a perfect system of conduction.* A ship having three masts, only one of which was protected, had her unprotected masts shattered, while the protected one remained untouched.

230. *Why should the conductors be attached to the external metal work of the building?*

Because the metal pipes, &c., would aid the conductors to convey the electricity away.

231. *Why should the conductors terminate in the earth?*

Because then the electricity would be discharged from the conductors, and would have a *sufficient area of escape,* without creating a shock. It is a good plan to attach the conductors to the gas-pipes in the earth, or to convey them to any body of water near the building.

232 *Is it dangerous to stand near a lightning conductor?*

It is *not* dangerous, because metal, being a better conductor than the human body, electricity would not leave the metal to pass through a worse conductor.

"God thundereth marvellously with his voice: great things doeth he, which we cannot comprehend."—JOB XXXVI.

233. *When is it dangerous to stand near an elevated object?*

When electric storms occur, it is dangerous to stand near tall objects which are *indifferent conductors*, because the fluids of the human body are a good medium, and lightning would pass through the body in preference to any worse mediums.

234. *What is the cause of the aurora borealis?*

The *aurora borealis* arises, in all probability, from the mingling of the electricities in the upper atmosphere. By some they are regarded as arising from the passage of electricity from the equator to the poles.

When they occur in the north, they are called *aurora borealis;* when in the south, *aurora australis.*

235. *What is the cause of aereolites, shooting stars, meteoric stones, &c.?*

Various opinions have been advanced respecting the origin of these bodies.

At one time they were believed to be projected from volcanoes in the moon. But it has more recently been ascertained that the lunar volcanoes are not now in an active state.

Then arose the opinion that they were aggregations of metallic vapours, which, meeting in the atmosphere, solidify there, and fall; just as watery vapours solidify and descend.

Another opinion (which appears less probable than the latter one) is, that they are planetary fragments, which revolve in orbits of their own; that our planet encounters periodic shoals of these little worlds, some of which become entangled in the earth's attraction, enter our atmosphere, and becoming heated to luminosity by friction against the atmosphere, are either consumed, as shooting stars and fire balls, or fall to the ground as aereolites.

236. *What is the cause of the Will-o'-the-Wisps, or Jack-o'-Lanterns?*

They are the result of marsh gas, or phosphuretted hydrogen gas, which emanates from decaying animal or vegetable substances, and which spontaneously ignite.

"All ye inhabitants of the world and dwellers on the earth, see ye, when he lifteth up an ensign on the mountains; and when he bloweth a trumpet, hear ye."—ISAIAH XVIII.

CHAPTER X.

237. *What is magnetism?*

Magnetism is the electricity of the earth, and is characterised by the circulation of *currents of electricity passing through the earth's surface.*

238. *What are magnetic bodies?*

Magnetic bodies are those that exhibit phenomena which show that they are under the influence of *terrestrial electricity,* and which indicate the direction of the *poles,* or *extreme points, of magnetic force.*

239. *What is Galvanism?*

Galvanism is the action of *electricity upon animal bodies,* and is so called from the name of its first discoverer, Galvani.

240. *What is Voltaic electricity?*

Voltaic electricity is the electricity that is developed during *chemical changes,* and is so called after Volta, who enlarged upon the theory of Galvani.

241. *What are the differences between mechanical, or frictional electricity, Voltaic electricity, Galvanism, and magnetism?*

Frictional electricity is electricity *suddenly* liberated under the effects of the *motion,* or the mechanical disturbance of bodies.

Voltaic electricity is a *steady flow* of an electric current, arising from the *gradual changes of chemical* action.

Galvanism and *Voltaism* are also *identical,* since the latter is founded upon, and is a development of, the former. But the term *Galvanism* is frequently used when speaking of the development of electricity in *animal bodies.*

Magnetism is the electricity of the *earth,* and is also understood to include the *fixed electricity of terrestrial bodies.*

242. Man knows not *what electricity is;* yet, by an attentive observance of its *effects,* he avails himself of the power existing in an unknown source, and produces marvellous results. When the Grecian philosopher, Thales, sat rubbing a piece of amber, and watching the attraction of small particles of matter to its

"To him that rideth upon the heavens of heavens, which were of old; lo, he doth send out his voice, and that a mighty voice."—PSALM LXVIII.

surface, he little knew of the mighty power that was then whispering to him its offer to serve mankind. And when Franklin, with the aid of a boy's plaything, drew down an electric current from the clouds, and caught a spark upon the knuckles of his hand, *even he* little conjectured that the time was so near when that strange element, which sent its messenger to him along the string of a kite, would become one of man's most submissive servants.

243. So many great results have sprung from the careful observation of the simplest phenomena, that we should never pass over inattentively the most trifling thing that offers itself to our examination. Nature, in her revelations, never seeks to *startle* mankind. The formation of a rock, and the elaboration of a truth, are alike the work of ages. It was the simple blackening of silver by the sun's rays which led to the discovery of the *chemical agency of light.* It was the falling of an apple which pointed Newton to the discovery of the *laws* of *gravitation.* It was the force of steam, observed as it issued from beneath the lid of a kettle, that led to the invention of the *steam-machine.* And it is said of Jacquard, that he *invented the loom* which so materially aided the commerce of nations, while watching the motions of his *wife's fingers*, as she plied her knitting. As great discoveries spring from such small beginnings, who among us may not be the herald of some great truth—the founder of some world-wide benefaction?

244. That the area of discovery has not perceptibly narrowed its limits, is evident from the fact that the greatest elements in nature are still mysteries to man. And though it may not be within the power of a finite being to unravel the chain of wonders that enfold the works of an infinite God—still it is evident, from the progress which discovery has made, and from the good which discovery has done, that God *does* invite and encourage the human mind to contemplate the workings of Divine power, and to pursue its manifestations in every element, and in every direction.

245. The wonderful force of *electricity* astonishes us the more when we view it in contrast with that equally wonderful element, *light.* We have seen that light travels with a velocity of 192,000 miles in a second, but that it falls upon a delicate balance so gently, that it produces no perceptible effect. As far as we know the nature of *electricity*, it is even *more ethereal* than *light;* yet, while the *ether of light* falls harmlessly and imperceptibly—even with the momentum of a flight of *ninety-five millions of miles*, the *ether of electricity*, bursting from a cloud only *five hundred yards* distant, will split massive stones, level tall towers with the dust, strike majestic trees to the ground, and instantly extinguish the life of man! *Why* does *the one ether* come divested of all mechanical force, while that which seems to be *even more ethereal* than it, is capable of exerting the mightiest force over material things? Does it not appear that the Creator of the universe has established these paradoxes of power to testify his Omnipotence—to show to man that with Him all things are possible; and that, in the grand cosm cism of the universe, every attribute of Omnipotence has been fulfilled?

246. Let us now consider man's relation to this Omnipotence. He sees that electricity smites the tall edifice, and observes that in doing so it displays choice for a certain substance through which it passes harmlessly, and that its violence is manifested only when its path is interrupted. Man, taking advantage of this preference of electricity for a particular conductor, stretches out an arm of that substance, and points it upwards to the clouds; electricity accepts the

"As for the earth, out of it cometh bread; and under it is turned up as it were fire."—JOB XXVIII.

invitation, and passes harmlessly to the earth. But this is not all; man learns by observation that electricity resides in all matter; that it may be collected or dispersed; that it travels along a good conductor at the rate of *half-a-million of miles in a second of time;* he constructs a battery, a kind of scientific fortress, in which he encamps the great warrior of nature; and then, laying down a conducting wire, he liberates the mighty force: but its flight must be on the path which man has defined, and its journey must cease at the terminus which he has decreed, where, by a simple contrivance of his ingenuity (the movements of a magnetic needle), the electric current is made to deliver whatever message of importance he desires to convey. Thus, the element which in an instant might deprive man of life, is subdued by him, and made the obedient messenger of his will!

247. *Why does the magnetic needle (the needle of the mariner's compass) point towards the north?*

The *cause* is not understood; but it is believed that currents of electricity are always passing through the earth's surface; and that these electrical currents originate *in the daily heating of the earth by the sun's rays,* in the direction of east to west.

It has already been explained, that any variation of temperature developes a current of electricity in the bodies that undergo the change. (178.)

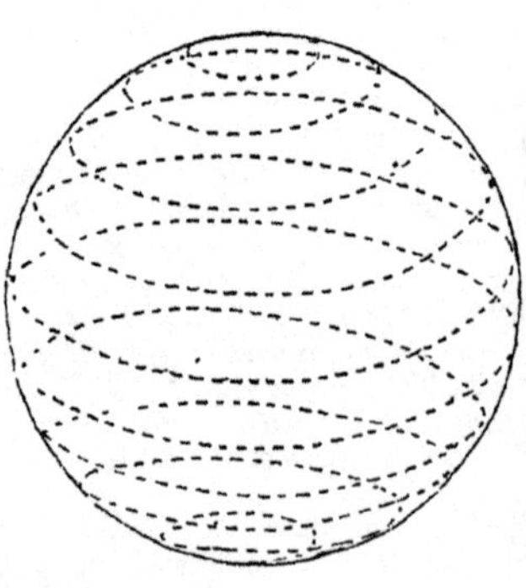

DIAGRAM ILLUSTRATING THE ELECTRICAL CURRENTS IN THE EARTH.

248. If a hollow paper globe be lined with a revolving copper wire, and a current of electricity be directed through the wire; a freely poised magnetic needle being presented to it will arrange itself in the direction of the poles of the globe, the points of the needle assuming the direction of right angles to the current of electricity. This illustration is supposed to accord with the phenomena of the earth's electricity, and the tendency of the magnetic needle to point to the north when the earth's electrical influence over it is undisturbed by extraordinary causes. It will be seen, therefore, that the *reason why* the needle points to the north is, that the currents of electricity passing through the earth's surface, give to magnetic bodies that are free to move, the tendency to arrange themselves in the direction of the north and south poles of the earth. But *why* electricity exercises this influence over magnetic bodies is unknown.

"Canst thou send lightnings, that they may go, and say unto thee, Here we are?"—JOB XXXVIII.

249. *Why can this tendency of the magnetic needle be made available for the transmission of signals?*

Because, whenever a magnetic needle is brought under the influence of an artificial current of electricity, the influence of the earth's electricity over the needle is *temporarily overpowered,* and the needle arranges itself at right angles with the current passing near to it. This tendency in the needle is called its *deflection.*

Deflection—the act of turning aside.

250. *How can this deflection of the needle be made to communicate telegraphic signals?*

It is only necessary to agree to the *meanings* of a certain *set of signals* made by the movements of the needle, and then any kind of information may be conveyed by it.

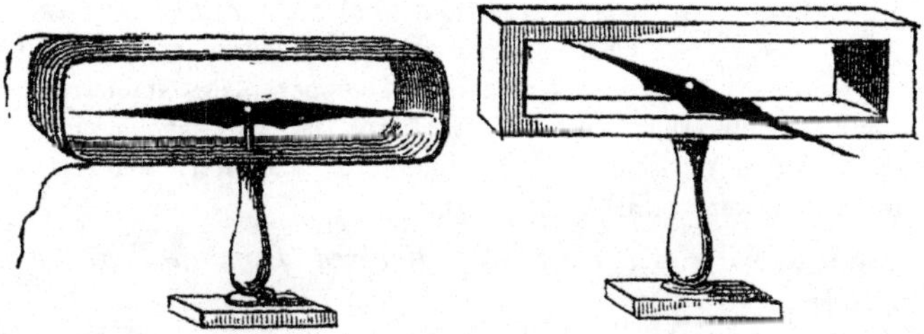

Fig. A. Fig. B.

251. Let Fig. A represent a magnetic needle freely suspended, and influenced only by the earth's electricity, which determines its direction, and causes it to point from south to north, or in the direction of the earth's poles. Let B represent the same needle, but let it be understood that a current of electricity is passing through a coil of wire surrounding the frame upon which the needle is suspended; the needle turns from its previous direction, and presents itself *at right angles to the current of electricity passing through the wire.*

252. *Why does the telegraphic needle turn from side to side, and strike against the small pins on either side of it?*

Because, while working, it is under the influence of artificial electricity; it turns from left to right, or from right to left, or assumes a state of rest, in obedience to the nature, the force, or the suspension of the electrical currents, which are under the

"Where wast thou when I laid the foundations of the earth? declare if thou hast understanding.

TELEGRAPHIC NEEDLE.

control of *the person who works the telegraph*, and who has the power of instantly turning on, or off, reversing, modifying, or strengthening the currents, by the aid of an ingenious machine.

253. *Why does the telegraphic needle rest in a vertical position, when its natural tendency is to point to the north?*

Because, as the upright position is the best for telegraphic purposes, a slight preponderance of weight is given to the south pole of the needle, and this causes it to lie in a vertical plane.

254. *Whence is the electricity derived which is directed along the wires?*

It is derived from what are termed *batteries*, in which a powerful chemical excitement is kept up by means of zinc and copper plates immersed in troughs where they are submitted to the corrosive action of diluted sulphuric acid. It is therefore *electricity evolved by chemical action*. (See 240.)

255. *How is the electricity directed from the battery to the wires?*

Simply by connecting, by the aid of the telegraphic machine, the ends of *the telegraph wires with the ends of wires communicating with the battery.*

256. *Why does electricity dart along the wires immediately the contact is made?*

Because the electricity evolved in the battery *seeks to escape*, and the moment a *conducting medium* is presented to it, it flies off with inconceivable rapidity along the surface of the conductor.

257. *What becomes of the electricity that passes along the wires?*

Wherever a wire begins or terminates, it is connected *with the earth*, by being attached to a block of metal, which is buried in the earth.

258. *What becomes of the electricity when it passes to the earth?*

It passes through the earth as instantaneously as it passed along the wire, *and returns to the battery again.* No matter, whether the distance be one mile or one thousand, *a perfect electric circle is formed,* and the electricity returns through the earth as certainly and as rapidly as it would have done, were a second wire laid down to conduct it back again.

The end of the battery from which the electricity escapes is called the *positive pole,* and that at which it returns is called the *negative pole.*

259. *What occurs as the electricity passes along the wires?*

It *deflects all the needles* attached to the telegraph-dials that are connected with the electric circle then in action.

260. *Why do the needles sometimes turn to the right, and at others to the left?*

Because the operator, who telegraphs the message, has the power of connecting the wires with either the *negative* or the *positive* currents. The change of the current *reverses the position of the needle,* and thus a greater number of movements, which can be arranged into signals, have been obtained.

261. *What is the velocity of electricity when thus transmitted?*

It is calculated that if a perfect system of wires could be provided, a despatch might be sent *many times round the world during two beats of a clock.*

262. *Why are the wires suspended upon posts?*

Because they are thereby isolated from conducting bodies, which would *bear the electricity away,* and destroy the effect of the telegraph.

263. *Why are the wires at the posts passed through bells, or nobs of earthenware or glass?*

Because, being non-conductors, those bodies, while they hold the wires firmly in their places, prevent the electricity *from passing through the wood of the posts to the earth.*

"Whereupon are the foundations thereof fastened? or who laid the corner-stone thereof?"—JOB XXXVIII.

264. *Why are wires that are laid down under ground, or through water, coated with gutta percha?*

Because gutta percha, is a *non-conductor*, and prevents the electricity from passing into the earth or the water.

CHAPTER XI.

265. *What is attraction?*

Attraction is the tendency of bodies to *draw near to each other*. It is called *attraction*, from two Latin words signifying *drawing towards*.

266. *How many kinds of attraction are there?*

There are five principal kinds of *attraction*:—

1. The attraction of *gravitation*.
2. The attraction of *cohesion*.
3. The attraction of *chemical affinity*.
4. The attraction of *electricity* and of *magnetism*.
5. And *capillary attraction*.

267. *Why do all bodies heavier than the air fall to the earth?*

Because they are influenced by the *attraction of gravitation*, by which all bodies are drawn towards *the centre of the earth*.

268. *Why do bodies lighter than the air ascend?*

Because the air, being a denser body, *obeys the law of attraction*, and in doing so *displaces lighter bodies* that interfere with its gravitation.

269. *Why do fragments of tea, and bubbles floating upon the surface of tea, draw towards each other, and attach themselves to the sides of the cup?*

Because they are influenced by the attraction of *cohesion*.

Cohesion—the act of sticking together.

"Remember that these magnify his work, which men behold. Every man may see it; man may behold it afar off."—JOB XXXVI.

270. *Why will a drop of water upon the blade of a knife leave a dark spot?*

Because the *iron of the knife attracts the oxygen of the water* by *chemical affinity;* and the two substances form a thin coating of *oxide of iron.*

Affinity—attraction between dissimilar particles through which they form new compounds.

271. *Why do clouds move sometimes towards each other from opposite directions? and*

272. *Why do light particles of matter attach themselves to sealing wax, excited by friction?*

Because they are moved by the *attraction of electricity.*

273. *Why will a towel, the corner of which is dipped in water, become wet for above the water?*

Because a water is conveyed up through the towel, by *capillary attraction.* The atoms of the water are attracted by the *threads of the towel,* and drawn up into the *small spaces between the threads.*

Capillary—resembling a hair, small in diameter.

274. *Why do small bodies floating upon water move towards larger ones?*

Because the attractive power of *a large body* is greater than that of *a small one.* As each atom of matter has inherent power of attraction, it follows that a *large aggregation of particles* must attract in proportion to the number of those particles.

275. *Why do clouds gather around mountain tops?*

Because they are *attracted by the mountains.*

276. *Why would a piece of lead tied to a string, and let down from a church steeple, incline a little from the perpendicular towards the church?*

Because the *masses of stone* of which the church is built would *attract the lead.*

"Who hath measured the waters in the hollow of his hand, and meted out heaven with the span, and comprehended the dust of the earth in a measure, and weighed the mountains in scales, and the hills in a balance?"—ISAIAH XL.

277. *How can man weigh the earth?*

By observing what is called the *deflection* of small bodies *when brought within given distances of larger bodies*, the degree of attraction *exercised by the large body upon the smaller one* becomes known. The attraction of the *large body* exercised over the *smaller body* is an opposing influence, *acting against the earth's attraction of the small body*, which is drawn out of its course; it constitutes a *natural balance between the influence of the earth and another body, acting in opposition to it.* Founded upon these, and some other data, man can weigh the earth, and give a morally certain result!

Deflection—the act of turning aside.

278. *How can man weigh the planets?*

The planets exercise as certain an influence upon each other *as do two pieces of wood floating upon a basin of water.* As the planetary bodies fly through their prescribed orbits, *and approach nearer to, or travel further from, each other*, they are observed to *deviate* from that course which they must have pursued *but for the increase or the decrease of some influence of attraction.* By making observations *at various times*, and by comparing *a number of results*, it is possible *to weigh any planetary body, however vast, or however distant.*

279. *How can man measure the distances of the planets?*

By making observations at *different seasons of the year*, when the earth is in *opposite positions in her orbit;* and by recording by *instruments constructed with the greatest nicety, the angle of sight* at which the planetary body is viewed; by noticing, also, *the various eclipses*, and estimating *how long the first light after an eclipse has ceased* reaches the earth, it is possible to estimate the *distances* of heavenly bodies, *no matter how far in the depths of the universe those orbs may be.*

280. *What are the opinions founded upon estimates respecting the magnitude of the sun?*

The *diameter* of the *sun* is 770,800 geographical miles, or 112 times greater than the diameter of the earth; its *volume* is

1,407,124 times that of the earth, and 600 times greater than *all the planets together;* its *mass* is 359,531 times greater than the earth; and 738 times greater than that of *all the planets.* A *single spot* seen upon its surface has been estimated to extend over 77,000 miles in diameter, and a *cluster of spots* have been estimated to include an area of 3,780,000 miles.

281. *What is the weight of the earth?*

The earth has a *circumference* of 25,000 miles, and is estimated to *weigh* 1,256,195,670,000,000,000,000,000 tons.

282. *What is the specific gravity of a body?*

It is its weight estimated *relatively to the weights of other bodies.*

283. *What determines the force with which bodies fall to the earth?*

Generally speaking, their *specific gravity,* which is proportionate to the density, or *compactness of the atoms* of which they are composed.

284. *Why does a feather fall to the earth more gradually than a shilling?*

Because the *specific gravity* of the feather and of the shilling is *relative to that of the air,* the medium through which the feather and the shilling pass. If there were *no air,* a shilling and a feather dropped at the same time from a height of forty miles *would reach the earth at the same moment.*

285. *What is repulsion?*

Repulsion is that property in matter by which it *repels or recedes from,* those bodies for which it has *no attraction or affinity.*

286. *Why does dew form into round drops upon the leaves of plants?*

Because it *repels the air,* and the *substances of the leaves* upon which it rests. Because, also, its own particles *cohere.*

"But as it is written, Eye hath not seen, nor ear heard, neither have entered into the heart of man, the things which God hath prepared for them that love him."—I. CORINTH. II.

287. *Why do drops of water roll over dusty surfaces?*

Because they *repel* the particles of dust; and also because their own particles have *a stronger attraction for each other* than for the particles of dust.

288. *Why does a needle float when carefully laid upon the surface of water?*

Because the needle and the water *mutually repel each other.*

289. *Why does water, when dropped upon hot iron, move about in agitated globules?*

Because the *caloric* repels the particles of the water.

290. *Why does oil float upon the surface of water?*

Because, besides being specially lighter than water, the particles of the oil and water *mutually repel each other.*

CHAPTER XII.

291. *What is the atmosphere?*

The *atmosphere* is the transparent and elastic body of mixed gases and vapours which envelopes our globe, and which derives its name from Greek words, signifying *sphere* of *vapour.*

292. *To what height does the atmosphere extend?*

It is estimated to extend to from *forty to fifty miles* above the surface of the earth.

293. *Why is it supposed that the atmosphere does not extend beyond that height?*

Because it is found, by experiment and observation, that the air becomes *less dense* in proportion to its altitude from the earth's surface. The gradual decrease of atmospheric density observed in ascending a mountain, or in a balloon, supplies sufficient data to enable us to calculate the height at which the atmosphere would probably *altogether cease.*

"I therefore so run, not as uncertainly; so fight I, not as one that beateth the air."—CORINTH. IX.

294. *What is the amount of atmospheric pressure at the earth's surface?*

The pressure of the atmosphere at the earth's surface is *fifteen pounds* to every square inch of surface. That is to say, that the column of air, extending fifty miles over a square inch of the earth, presses upon that square inch with a weight equal to *fifteen pounds.*

295. *Is that the weight of dry or moist air?*

That is the weight of air at what is called the *point of saturation,* when it is fully charged with *watery vapour.*

296. *What is the proportion of watery vapour in the atmosphere?*

The proportion *constantly varies.* Evaporation is not a result of accident; it seems an *established law* that the air shall constantly *absorb vapour* until it has reached the maximum that it can hold. Experiments have been tried, in which dry air has been pressed upon the surface of water with great force, *but no degree of pressure could prevent the formation of vapour.*

297. *What is the total amount of atmospheric pressure on the earth's surface?*

The total amount of atmospheric pressure on the earth's surface, at 15lbs. to the square inch, amounts to 12,042,604,800,000,000,000lbs. This pressure is equal to that of a globe of lead of *sixty miles in diameter.*

298. *What is the pressure of the atmosphere upon the human body?*

Estimating the surface of man's body to be equal to *fifteen square feet,* he sustains an atmospheric pressure of 32,400lbs., or nearly *fourteen tons and a half.* The mere *variation of weight,* arising out of the changes in the state of the atmosphere, may amount to as much as a *ton and a half.*

299. *Why does man not feel this pressure?*

Because the diffusion of air which, *surrounding him in every direction,* and acting upon the *internal* as well as the *external*

"Ascribe ye strength unto God: his excellency is over Israel, and his strength is in the clouds."—PSALM LVIII.

surfaces of his body, and probably *surrounding every atom of his frame*, establishes an equilibrium, in which every degree of pressure *counteracts and sustains itself*.

300. *What is the weight of air relative to that of water?*

A cubic foot of air weighs only 523 grains, a little more than *an ounce*; a cubic foot of water weighs *one thousand ounces*.

301. *What is the greatest height in the atmosphere which any human being has ever reached?*

M. Gay Lussac, in the year 1804, ascended to the height of 23,000 feet.

302. At an altitude of 18,000 feet the air is indicated by the barometer to be only *half as dense* as at the surface of the earth. And as the densities of the atmosphere decrease in a geometrical progression. the density will be reduced to *one-fourth* at the height of 36,000 feet, and to *one-eighth* at 54,000 feet. The effects of the decreasing density of the atmosphere are, that the *intensity of light and sound are diminished, and the temperature is lowered*. Persons who have reached a very high elevation, state that the sky above them began to assume the appearance of darkness; and there can be no doubt that, if it were possible to reach an altitude of some fifty to sixty miles, there would be *perfect blackness, although the sun's rays might be pouring through the darkened space* to illuminate the atmosphere. Upon the summit of Mont Blanc, the report of a pistol at a short distance can *scarcely be heard*. When Gay Lussac reached the height of 23,000 feet, he breathed with great pain and difficulty, and felt distressing sensations in his ears, as though they were *about to burst*. Upon the high table-lands of Peru, the lips of Dr. Ischudi cracked; and blood flowed from his eyelids.

303. *What is a vacuum?*

A vacuum is a space *devoid of matter*. The term is generally applied to those instances in which air is drawn from within an air-tight vessel.

304. *Is it possible to form a perfect vacuum?*

It is probably *impossible to do so*, even with the most powerful instruments—some portion of air would remain, but in so thin a form that it would be *imperceptible*.

305. *Why does the depression of a pump handle cause the water to flow?*

Because the putting down of the handle lifts up the piston with the *valve closed*, thereby tending to produce a *vacuum*; but *the pressure of the air* upon the water *not contained in the pump,* forces more water up into the part where a *vacuum* would otherwise be formed. Then, when the handle is raised, and the piston forced downwards, *the valve opens,* and the water rushes through.

There is a second valve below the piston, which closes with the downward movement, to prevent the water from *rushing back again.*

306. *How high will atmospheric pressure raise water in the bore of a pump?*

It will raise water to an elevation of *thirty feet* above its level.

307. *Why will it raise water to an elevation of thirty feet.*

Because a column of water of *thirty feet high,* nearly balances the weight of *a column of air* of equal surface, *extending to the whole height of the atmosphere.* When, therefore, water is elevated to the height of thirty feet, the power of the pump is enfeebled, as the air and the water *balance each other.*

308. *How is water raised to a greater elevation when it is required?*

By mechanical contrivances, by which the water is *forced* to a greater elevation.

309. *Why does water run through the bent tube called a syphon?*

Because the atmospheric pressure upon the water on *the outside of the syphon* forces it into the tube as fast as the syphon empties itself through its longer arm.

310. *Why does water run through the longer arm of the syphon?*

Because the weight of the water in the longer arm of the syphon *is greater than that in the shorter;* therefore it runs out by its own gravity. And, as in running out, it creates a tendency towards a *vacuum,* the pressure of the outer air comes into operation and forces the water through the tube.

"He stretcheth out the north over the empty place, and hangeth the earth upon nothing."—JOB XXIV.

311. *Why does water issue from the earth in springs?*

Some springs are caused by *natural syphons* formed in the fissures of rocks, which, communicating with bodies of water are continually filled by atmospheric pressure, and therefore convey streams of water to the point where they are set free.

312. *Why, if a wine glass be filled with water, a card laid upon it, and the whole inverted, will the water remain in the glass?*

Because the pressure of the atmosphere upon the surface of the card counteracts the weight of the water.

313. *What has the card to do with the experiment?*

It forms *a base* upon which the water may rest, while the glass is being inverted; and prevents the air from acting upon the *fluidity* of the water, and forcing it out of the glass.

314. *Why can flies walk on the ceiling?*

Because their feet are so formed that they can form a *vacuum* under them; their bodies are therefore sustained in opposition to gravitation by *atmospheric pressure.*

315. *How did Mr. Sands perform the feat of walking across the ceiling?*

By having large discs of wet leather attached to his feet, so that when they were placed upon a smooth surface, the air was excluded and when he allowed his weight to act upon one of the discs, it formed a *hollow cup* and a *vacuum.* By forming a vacuum of only *twelve square inches* he gained a pressure of 180lbs.; this being more than his weight, he could accomplish the feat with no other difficulty than that of remaining in an inverted position. The air was admitted underneath the discs by valves, which were closed by springs, which being pressed by the heels of the performer, let in the air, and *set the feet free.*

316. *Why is it difficult to strike limpets from rocks?*

Because they have the means of forming a *vacuum* under their shells, and are pressed on to the rocks by the weight of the atmosphere.

"For he looketh to the ends of the earth, and seeth under the whole heaven; to make the weight for the winds."—JOB XXVIII.

317. *Why can snails move over plants in an inverted position?*

Because they form a *vacuum* with the smooth and moist surfaces of their bodies, and are supported by atmospheric pressure.

CHAPTER XIII.

318. *What is wind?*

Wind is air *in motion.*

319. *What are the velocities of winds?*

A *breeze* travels ten feet in a second; a *light gale,* sixteen feet in a second; a *stiff gale,* twenty-four feet in a second; a *violent squall,* thirty-five feet in a second; *storm wind,* from forty-three to fifty-four in a second; *hurricane* of the temperate zone, sixty feet in a second; *hurricane* of the torrid zone, one hundred and twenty to three hundred feet in a second. When wind flies at one mile an hour, it is scarcely perceptible. When its velocity is one hundred miles an hour, it tears up trees, and devastates its track.

320. *What are trade winds?*

Trade winds are vast currents of air, which *sweep round the globe* over a belt of some 12,000 miles in width.

321. *What is the cause of trade winds?*

The air over the tropical regions becomes heated and ascends; it then diverges in two high currents, one towards the north, and the other towards the south pole, where, being cooled, it again descends, and returns towards the equator to replace the air as it ascends therefrom. There is, therefore, a constant revolution of vast currents of air between the tropics and the poles, producing *north and south winds.*

"And God made a wind to pass over the earth."—GENESIS VIII.

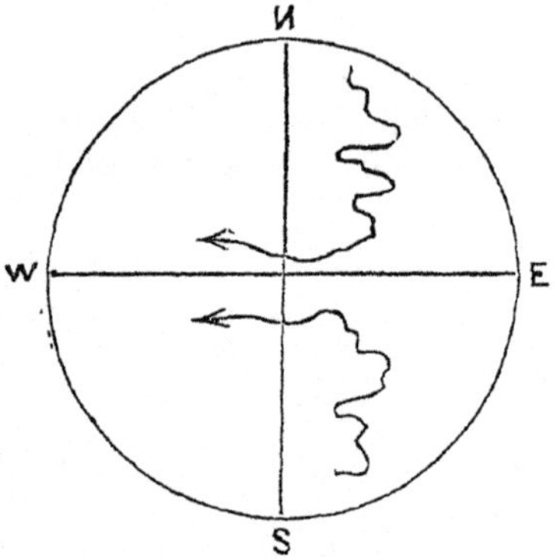

322. *Why do the trade winds blow from east to west, though, in their origin, their direction is from north to south, and from south to north?*

Because, as the north and south winds blow towards the equator, they are affected by the revolution of the earth from *west to east.* As the two winds from the poles approach the equator, they are gradually diverted from their northerly and southerly course, to an easterly direction, by the revolution of the earth.

323. *Why is there a prevalence of calms at the equator?*

Because, as the north and the south winds move towards the equator, they drive before them volumes of atmosphere, which, meeting in opposite directions, resist and counterpoise each other, and abide in a state of stillness between the north and south-easterly winds, one on the north and the other on the south of the equator.

324. *What are monsoons?*

Monsoons are *periodical winds* which blow at a given period of the year from one quarter of the compass, and in another period of the year from the opposite quarter of the compass.

325. *What is the cause of monsoons?*

Monsoons are caused by changes in the position of the sun. When the sun is in the southern hemisphere, it produces a *north-east wind,* and when it is in the northern hemisphere, a *north-west wind.* The north-east monsoon blows from November to March, and the south-west monsoon from the end of April to the middle of October. The region of monsoons lies a little to the north of the northern border of the trade wind, and they blow with the greatest force, and with most regularity, between the eastern coast of Africa and Hindostan.

"He shall blow upon them, and they shall wither, and the whirlwind shall take them away as stubble."—ISAIAH XI.

326. *What determines the character of winds?*

The character of winds is influenced by the condition of *the surfaces over which they blow.* Winds blowing over dry and arid plains and deserts are *dry and hot.* Winds blowing across snow-capped mountains and regions of ice are *cold.* Winds that cross oceans are *wet;* and those that cross extensive continents are *dry.*

327. *What winds are most prevalent in England?*

In England, out of a *thousand days,* north winds prevail in 82; north-east, 111; east, 99; south-east, 81; south, 111; south-west, 225; west, 171; north-west, 120.

328. *What is the cause of storms?*

Storms result from violent commotions of the atmosphere, and are chiefly the result of extreme *changes of temperature.*

The *magnetic* state of the earth, and the *electrical* state of the atmosphere, also materially influence the phenomena of storms.

By some persons the theory is entertained that storms result from various winds *rushing into a centre* in which the atmosphere has become extremely condensed. According to this theory, a storm is a mighty whirlwind.

329. A most violent hurricane occurred in 1780, which destroyed Lord Rodney's fleet, and a vast number of merchant ships. It is said to have killed 9,000 persons in Martinique alone, and 6,000 in St. Lucia. The town of St. Pierre, in Martinique, was totally destroyed; and only fourteen houses in the town of Kingston, in St. Vincent, were left uninjured.

330. *Why do the most violent storms occur in and near the tropics?*

Because there the temperature is very high, and the cold currents of air rushing towards the equator from the poles, causes great *atmospheric disturbance.*

331. *What are whirlwinds?*

Whirlwinds are produced by violent and contrary currents meeting and striking upon each other, producing *a circular motion.* They generally occur after long calms, attended by much heat.

"They shall be as the morning cloud, and as the early dew that passeth away, as the chaff that is driven with the whirlwind out of the floor, and as the smoke out of the chimney."—HOSEA XIII.

332. Whirlwinds occuring at sea, or over the surface of water, sometimes put the water in motion, and as the wind rises upwards it lifts with it a whirling mass of water, producing a *water spout.*

WATER SPOUT.

333. *Why do the wings of wind-mills turn round?*

Because the wind, striking *at an angle* upon the wings, forces them aside; and as there are four wings all upon the same angle, and fixed upon the same centre, the *oblique pressure* of the wind causes the centre to rotate.

334. There is a world of *miniature phenomena* which has never been fully recognised, in which we may see the mightier works of nature pleasingly and truthfully illustrated.

When the wind blows into the corner of a street, and whirling around, catches straw, dust, and feathers in its arms, and then wheels away, flinging the troubled atoms in all directions—it is a miniature of the mightier *whirlwind*, which wrecks ships, uproots trees, and levels houses with the earth.

When a cloud of dust, on a hot summer's day, rises and flies along the thirsty road, making the passenger close his eyelids, and dusting the leaves of wayside vegetation—it is a miniature of the terrible *simoom*, which blows from the desert sands, scattering death and devastation in its track.

When steam issues from the tea-urn, and becomes condensed in minute drops upon the window pane—the miniature is of the *earth's heat*, evaporating the waters, and the cold of night condensing the vapours into *dew*.

"The wind bloweth where it listeth, and thou hearest the sound thereof, but canst not tell whence it cometh, and whither it goeth; so is every one that is born of the Spirit."—II. JOHN III.

When grass and corn bend before the wind, and are beaten down by its force; when the pond forgets its calm, and rises in troubled waves, casting the flotilla of natural boats, that move upon its surface, in rude disorder upon its windward shore—the little storm is but a miniature of those great *hurricanes* which wrecked a fleet in the Black Sea, and levelled the encampments of a mighty army.

When the snow that has gathered upon the house-top, warming beneath the smiles of the sun, slips from its bed, and drops in accumulated heaps from the roof—it is a miniature of those terrible *avalanches* which in the Pyrenees bury villages in their icy pall, and doom man and beast to death.

When the rivulet hurries on its course, and meeting with obstructions, leaps over them in mimic wrath, overturning some little raft upon which, perchance, a weary fly has alighted—it is a miniature of those *rapids* on whose banks the hippopotamus and the alligator yet live; and where, though rarely, man may be seen directing his raft over the troubled current, amid the rush of *debris* from forests unexplored.

And when, in a basin of the rivulet, two opposing currents meet, and form a little vortex into which insect life and vegetable fragments coming within the sphere of its influence are drawn—it is a miniature of the roaring *whirlpool*, or the wilder *maelstrom* of the Norwegian seas.

Nature rehearses all her parts in mild whispers; for every picture that she paints, she places a first study upon the canvas. Man need not go into the heart of her terrors to understand their laws. Many an unknown Humboldt, sitting by the river's side, may rejoice in the "aspects of nature," and share the bliss of knowledge with the great philosopher.

CHAPTER XIV.

335. *What causes the rainbow?*

The *refraction* of the sun's rays by the falling rain.

336. *Why are the colours of the rainbow called prismatic?*

Because they are produced by the refraction of the rays of light by the rain, in the same manner as they are refracted by the *prism*.

337. *Why do films of oil, floating upon water, display prismatic colours?*

Because the varying thickness of the films produces prismatic refraction of light.

338. *Why do mother of pearl, the feathers of birds, the rings in rock crystal, &c., &c., exhibit such beautiful tints?*

Because the peculiar condition of their surfaces refracts the rays of light, unfolding as it were the coloured rays of the pencil of light.

339. *Why are there sometimes two rainbows?*

Because a portion of the light refracted by the rain which causes the primary bow is *reflected,* and falls upon other rain drops.

340. *Why is the secondary bow fainter than the primary bow?*

Because it is the result of light twice decomposed, whereas the true rainbow is the result of light only once decomposed. The secondary rainbow is also produced by drops of water very far off.

341. *Why is the order of the colours of the secondary bow reversed?*

Because it is a reflection of the primary bow, and, like all reflections, exhibits an arrangement precisely the opposite of that of the object reflected.

342. *What are lunar rainbows?*

Lunar rainbows are caused by the light of the moon falling upon rain. They are exceedingly rare, and are seldom coloured.

343. *Why are lunar rainbows more frequent in northern latitudes than in our climate?*

Because there the moon shines with a brilliancy unknown to us, consequently *lunar rainbows* are more frequently produced, and are occasionally coloured.

344. *What is the cause of the optical illusions frequently observed in nature?*

There are various kinds of natural optical illusions :—

The *mirage,* in which landscapes are seen reflected in burning sands.

The *fata morgana,* in which two or three reflections of objects occur at the same time.

"And it shall come to pass, when I bring a cloud over the earth, that the bow shall be seen in the clouds."—GENESIS IX.

The *ærial spectra,* or reflections in the atmosphere of terrestrial objects, &c

ILLUSTRATING THE APPEARANCE OF PHANTOM SHIPS.

The optical illusions above enumerated owe their origin to various atmospheric conditions, in which *refractions* and *reflections* are *multiplied* by the different densities of atmospheric layers. They chiefly occur in hot countries, where, from the varying effects of heat, the conditions of atmospheric refraction and reflection frequently prevail in their highest degree.

345. *What are haloes?*

Haloes are *luminous circles* occasionally seen round light-giving bodies, more particularly the sun and moon; they chiefly arise from the *refraction of light by vaporous water,* either in the form of clouds, or in a state of diffusion through the atmosphere.

"When a halo girts the moon,
Rain may be expected soon."

346. *Why do haloes foretell that rain may be expected soon?*

Because they are produced by the *water suspended in the air.*

"Thou, O God, didst send a plentiful rain, whereby thou didst confirm thine inheritance, when it was weary."—PSALM LXVIII.

347. *Why do large haloes indicate the more speedy approach of rain than small ones?*

Because the largeness of the halo indicates that the clouds are *near the earth*, and that therefore they are likely to discharge rain.

"A rainbow in the morning is the shepherd's warning;
A rainbow at night is the shepherd's delight."

348. *Why is "a rainbow in the morning the shepherd's warning?"*

Because a rainbow cannot occur unless when the rain is falling *opposite to the sun;* in the morning the rainbow appears *in the west*, or rainy quarter, and therefore there is a probability of a *duration* of rain.

349. *Why is "a rainbow at night the shepherd's delight?"*

Because a rainbow at night (or, more properly, in the evening) occurs *in the east*, which, being a dry quarter, indicates the probability that the rain *will not continue long.*

350. *Why do figures viewed through the hot air proceeding from furnaces, and from lime-kilns, appear distorted and tremulous?*

Because the ever-varying density of the air which flies away in hot currents, and is succeeded by cold, *constantly changes the refractive power* of the medium through which the figures are viewed.

351. *Why do stars twinkle?*

Because their light reaches us through *variously-heated and moving currents of air.* In this case the earth is the *kiln*, and the *stars* the *object* that is *viewed through the refractive medium.*

352. *Why does much twinkling of the stars foretell bad weather?*

Because it denotes that there are *various aerial currents* of different temperatures and densities, producing *atmospheric disturbance.*

"Is not God in the height of the heaven? and behold the height of the stars, how high they are!"—JOB XII.

353. *Why does twilight occur before sun-rise and after sun-set?*

Because of the refractive effects of the atmosphere. Rays of light, passing obliquely from the sun through the air to the earth, are refracted three or four times by the varying density of the medium. Each refraction bends the rays towards the *perpendicular;* and hence we see the sun *before it rises* and *after it sets.*

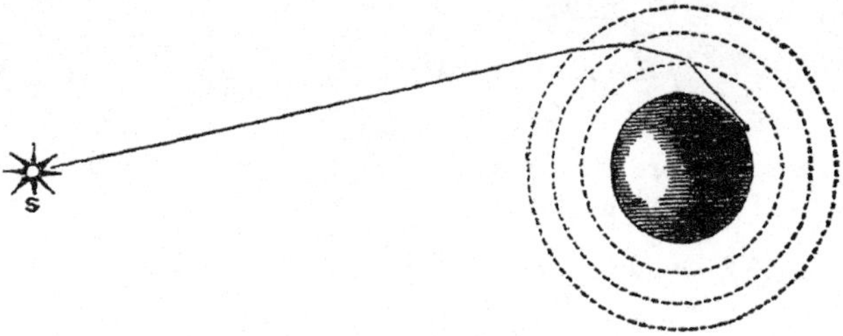

DIAGRAM EXHIBITING THE REFRACTION OF THE SUN'S RAYS IN PASSING THROUGH THE ATMOSPHERE.

CHAPTER XV.

354. *What is rain?*

Rain is the result of the condensation of the *vapours forming the clouds,* which, as drops of water, fall to the earth.

355. *Why does the vapour of the clouds condense?*

The most obvious cause is the effect of *cold winds* or currents, acting upon the warmer vapour, and resolving it into drops of water.

356. *When is a cold current of air likely to produce rain?*

When the atmosphere is fully charged with vapour; that is, when it has reached the *point of saturation.*

"When ye see a cloud rise out of the west straightway ye say, There cometh a shower and so it is. And when ye see the south wind blow ye say, There will be heat; and it cometh to pass."—LUKE XII.

357. *What is the point of saturation?*

It is that state of the air when it has taken up *as much vapour as it can sustain.*

358. *In what seasons of the year does evaporation chiefly take place?*

In *spring* and *summer.* The relative amount of evaporation for the four seasons may be thus stated: spring, 8 inches; summer, 11; autumn, 6; winter, 3.

359. *Why is there greater evaporation in spring and summer than in autumn and winter?*

Because the *temperatures* of spring and summer are higher than those of autumn and winter.

360. *Why are drops of rain that fall in low districts larger than those that fall upon elevated places?*

Because the drops meeting with each other in their descent, *unite and form larger drops,* and they also unite with the vapour ascending from the earth.

361. *From what atmospheric heights does rain descend?*

The *cirrus* clouds, which are the highest, are said to float at a height of 24,000 feet, At this elevation the clouds must exist in the state of a very fine snow; the other clouds vary in their altitudes from 20,000 feet to 10,000, 5,000, &c., down to the *stratus* and the *nimbus,* the lowest clouds, which frequently touch the earth. The most frequent rain-forming altitude is probably from 5,000 to 10,000 feet.

362. The estimated heights given must be looked upon as very conjectural, although they have been derived from the best existing authorities. It is sufficient to know that the range of the altitude of the various clouds is from that of the *Nimbus,* or *thunder cloud,* 1,500 feet, to that of the *Cirrus,* 24,000 feet, the others being intermediate. The first three of the clouds, enumerated at 483, constitute what are called the *primary forms.* The remaining four are called *secondary forms,* because they arise, as their names generally indicate out of combinations of the *primary forms.* Although, from the frequent mingling of clouds, it is not always practicable to identify them by the adopted classification, still, as there is generally a prevalence of one type of cloud over

another, the observer will be able to distinguish a "*Cirrus-sky*," or "*Cirro-cumulus sky*," &c. Upon some occasions the typical characters of the clouds are beautifully defined; and the contemplation of their forms, and the laws of their formation, affords infinite pleasure to the observer. The advantages of scientific knowledge are such, that whether you look downwards to the earth, or upwards to the sky, you have still the writing of God to read.

363. *Why are showers, when the rain-drops are large, unlikely to continue?*

Because the largeness of the drops show that the clouds are flying at *a considerable height*, and it is therefore probable that a cold current of air has suddenly condensed their vapour. The cause being temporary, the effect will not continue long.

364. *Why does rain sometimes "drizzle?"*

Because the clouds producing the rain are *very near to the earth*, so that the drops have not time to *unite* in their short descent.

365. *Why are drizzling rains likely to continue long?*

Because they proceed from clouds *flying near to the earth*, and they therefore show the probability that the air is at the point of saturation.

366. *Why are mountainous districts rainy?*

Because the mountains *attract the clouds*, and the summits of mountains being *usually cold*, contribute to the condensation of the vapours.

367. *Why do winds influence the fall of rain?*

Because if they blow from a *dry quarter*, they disperse the vapour of the clouds; if they blow from a *wet quarter*, they add vapour to that already existing; if they blow from a *cold quarter*, they condense the vapours; if they blow from a *warm and dry quarter*, they disperse and rarefy the vapours of the clouds; but if they blow from a *warm country*, and cross the ocean, they then become loaded with moisture, which they discharge upon reaching a colder latitude.

"The south wind always brings wet weather,
The north wind wet and cold together;
The west wind always brings us rain;
The east wind blows it back again."

"Out of the south cometh the whirlwind; and cold out of the north."—
JOB XXXVII.

368. *Why does the south wind bring wet weather?*

Because it blows across the *South Atlantic ocean*, and being warmed by the tropical heat, takes up a great amount of moisture. which it precipitates as it becomes cooler.

369. *Why does the north wind bring wet and cold together?*

Because it comes over the cold and frozen regions of the *Arctic ocean*, deriving moisture from the ice and snow over which it blows, and when it mingles with and cools the air of our latitude, the amount of vapour is too great to be held in suspension, and is precipitated in cold and uncongenial rains.

370. *Why does the west wind bring us rain?*

Because it comes from *warm regions* across the *Great Atlantic*, out of whose bosom it drinks the vapour which it afterwards pours down in refreshing showers.

371. *Why does the east wind "blow it back again?"*

The east winds come to us across the *vast plains* of *Northern Germany*. In the latter part of the spring those plains are exceedingly cold, and the east winds are found to be very severe. Although cold, they do not bring rain, because their track has been over a vast area of land.

CHAPTER XVI.

372. *Why are drops of rain spherical?*

Because the particles of water of which they are composed are held together by *cohesion;* they attract each other, and thus arrange themselves around a common centre.

373. If a small stick be dipped in water, or any other fluid, and drawn out again, a drop will be found hanging at the end of it in a spherical form. The drop is spherical, because each particle of it exerts an equal force in every direction, drawing other particles towards it on every side so far as its influence extends. Thus the very formation of drops obviously demonstrates that there must exist a cause which produces this effect. This cannot be *gravity*, for that is rather an obstacle to the formation of drops; since by the weight of the particles, large globules resting on solid bodies are scattered, and their

regular spherical form prevented. To explain this phenomenon, there remains only the power of attraction, acting between the particles of the liquid body, for if it be supposed that the particles of a substance reciprocally attract each other with equal force, and their aptitude for being moved upon one another be great enough to overcome any impediment to their motion, it follows, by the principles of mechanics, that the equilibrium of the attractive forces only takes place when the mass has received a globular form. Hence it is that all liquid bodies assume a spherical form when suffered to fall through the air.

374. *Why are shot manufactories furnished with very high towers?*

Because shot are made upon the same principle as that which forms *spherical drops of rain:* the melted lead is scattered from the interior of the tower at its summit, and as the liquid metal falls it forms into spheres, cools, and hardens, forming round shot.

375. *Why do large drops of rain, when they strike the earth, become scattered in smaller drops?*

Because the force with which the large drops strike the earth, after falling from a great height, is sufficient to overcome *the cohesion of their particles.* But when that force is expended, the scattered drops again immediately form spheres, by the *cohesion* of their atoms.

376. *Why do drops of rain unite in little streams, and then in larger ones, and run downwards?*

Because, wherever drops of water meet, they immediately unite by *cohesion;* and they also obey the law of *gravitation,* which draws them towards the earth. They therefore run through any downward channel until they find a level from which they cannot escape.

377. *Why is rain water nearly pure?*

Because, being distilled by evaporation from the waters of the earth, it is *freed from every kind of animal or mineral matter,* by which the water from which it was derived might have been contaminated.

"If the sun in red should set,
The next day will be free from wet,
But if the sun should set in grey,
The next will be a rainy day."

"In the morning ye say, It will be foul weather to-day, for the sky is red and lowering."—MATT. XVI.

378. *Why "if the sun in red should set," will "the next day be free from wet?"*

Because the amount of vapour floating in the air *refracts the beams of the sun,* and more freely transmits the red rays of light than the other coloured rays. It is the *degree* of moisture in the atmosphere which affects the refraction of the light; and when the red rays of evening are freely transmitted, the amount of moisture does not approach the rain point.

See the Chapters on Light.

379. *Why "if the sun should set in grey," will "the next be a rainy day?"*

Because the greyness of the clouds denotes that the atmosphere, though warm with the rays of sunlight, is charged with vapour beyond that degree which transmits the red rays of light, and therefore the excess of humidity will probably soon fall as rain.

"An evening red, and morning grey,
Will set the traveller on his way;
But an evening grey, and a morning red,
Will pour down rain on the traveller's head."

380. *Why will "an evening red and morning grey, set the traveller on his way?"*

Because the redness of the sun-set indicated that the atmosphere was *charged only with light clouds,* and that the cold of night has condensed those clouds, causing them to appear grey; still, *when the warmth of the sun strikes upon them,* they will be dispersed, and no rain will fall.

381. *Why will "the evening grey and morning red, pour down rain on the traveller's head?"*

Because the greyness of the evening showed the air in the west to be *fully charged with moisture,* and as the sun rises in the east displaying a red sky, it shows that the humidity extends *also to the east, or dry quarter,* and that therefore rain may be expected.

These explanations of old weather proverbs are liable to certain modifications, as are the proverbs themselves. The direction of the wind, its temperature, the elevation of the sun, &c., &c., will necessarily be found to affect the prognostics of a morning or evening sky.

"When it is evening ye say, It will be fair weather, for the sky is red."—
MATT. XVI.

382. *What are mists?*

Mists are formed by the condensation of invisible vapour into minute drops of water near the earth's surface. They may be regarded as *unprecipitated dew.*

See Dew and the Radiation of Heat.

383. *Why do mists sometimes appear to float a short distance from the ground?*

Because the earth is then *warmer* than the stratum of air which lies immediately above it.

384. *Why do we frequently see what are called "creeping mists," while the air above them is quite clear, and the ground comparatively free from dew?*

Because during the night the earth commences *the radiation of its heat*, and cools the air lying immediately over its surface. But the air being calm, the condensed vapour, instead of passing away, and allowing the radiation of the earth to continue, whereby it would be further cooled, and the moisture precipitated as dew, forms a *stationary stratum of cloud* over the ground, preventing its further loss of heat. The mist, therefore, becomes poised between the still superincumbent air, and the drier air upon the earth's surface.

385. *Why, as the sun rises, do the mists disappear?*

Because the warmth of the sun again affects the temperature of the atmosphere, and the vapours which had been rendered visible by condensation again *become rarefied.*

386. *Why do mists continue to appear and disappear several days in succession?*

Their continuance depends upon the *duration of calm weather*, and the same alternations of temperature. As long as a still atmosphere prevails, with similar variations from the warmth of day to the cold of night, mists continue. But when winds arise, and the accumulating clouds in the higher atmosphere check the formation of dew and vapour near the earth, and gradually gather for the production of rain, then the mists cease.

"For the Lord said unto me, I will take my rest, and I will consider in my dwelling-place like a clear heat upon herbs, and like a cloud of dew in the heat of harvest."—ISAIAH XVIII.

387. *Why do mists indicate the continuing of rain?*

Because they show that *the air is charged with moisture*, which, upon a change of temperature, will be precipitated.

388. *Why are mists frequent in the neighbourhood of the sea, and near marshy places?*

Because the temperature of water is frequently higher than that of the surrounding air. The water therefore evaporates, but when it rises on the air it is *immediately condensed, and forms mists.*

389. *Why is water frequently warmer than the surrounding air?*

Because water *takes up the warmth of the earth*, and it does not radiate heat freely from its surface.

CHAPTER XVII.

390. *What is snow?*

Snow is the watery vapour of the clouds *gradually frozen*, its particles being arranged in crystalline forms, similar to those of hoar frost.

391. *What is meant by the snow line?*

The *snow line* is the estimated altitude in *all countries* where *snow would be formed.* Even at the equator, at an altitude of 11,000 to 12,000 feet from the level of the sea, snow is found upon the mountain summits, where it perpetually lies. As we proceed north or south from the equator the *snow line lessens in altitude.* Had we in England a mountain 12,000 feet high, it would be perpetually *crowned with snow.*

392. *What is red snow?*

Red snow is the name given to the snow in the arctic regions upon which a minute vegetable (probably the *Protoccus nivalis*)

"He giveth snow like wool: He scattereth the hoar frost like ashes."—
PSALM CXLVII.

grows, imparting to the snow a red colour. Recent microscopic investigations have shown it to consist of a minute vegetable cell which secretes a red colouring matter.

CRYSTALS OF SNOW, AS SEEN THROUGH A MICROSCOPE.

393. *How may the crystalline forms of snow be easily observed?*

By catching them as they fall upon the sleeve of a *black* mantle, or the crown of a hat. Their curious structure may then be seen.

394. *Why does snow keep the earth warm?*

Because it forms a *woolly covering* of a nature that will not allow the warmth of the earth to pass away; nor will it suffer the cold frosts to penetrate the soil.

395. *Why cannot the warmth of the earth escape through the snow?*

Because snow is a *bad conductor*, just as a carpet, or blanket, or any kind of woollen texture. It has, in fact, been frequently called "the snow blanket."

396. *How do we know that snow keeps the earth warm?*

Because a *thermometer* placed in the earth, *underneath the snow*, has been found to stand several degrees higher than a thermometer

"Hast thou entered into the treasures of the snow? or hast thou seen the treasures of the hail?"—JOB XXXVIII.

also placed in the earth, but having the snow removed. If snow be swept away from the earth in frosty weather, the moist earth will be found to freeze to a considerable depth, while that which is covered with snow will be protected from such severe effects.

397. *What benefits result from the warmth of the earth being preserved by "the snow blanket?"*

It prevents seeds, bulbs, and roots from being destroyed by *severe frosts.* Tender blades of plants, that would be nipped by the frosts, continue to grow while covered with snow, and find protection in its warm covering.

398. Snow is found to be of greater importance to man than is generally supposed. But, although in this country we are enabled to recognise the hand of Providence in the gift, there are latitudes wherein the blessing thus conferred is more deeply felt. In such countries as Canada, Sweden, and Russia, the falling of snow is looked for with glad anticipations, quite equalling those which herald the "harvest-home" of England, or the "vintage" of France. No sooner is the ground covered with snow, than cranky old vehicles that had been jolting over rough roads, and sticking fast in deep ruts of mud, are wheeled aside, and swift sledges take their place. Towns distant from each other find an easy mode of communication; the markets are enlivened, and trade thrives. Snow supplies a kind of railroad, covering the entire face of the country, and sledges glide over it, almost with the speed of the locomotive.

399. *Why are snow flakes so light?*

Their lightness arises from the extent of their surface being so great in proportion to the very *small amount of substance* which they contain. It is from this peculiarity that snow clouds will float in the higher regions of the atmosphere. The beautiful *cirrus* clouds are probably all formed of snow, as they float far above the *snow line.*

400. *Has the congelation of snow ever been witnessed?*

Yes; there are several very interesting cases upon record. A company of persons were met in an assembly-room in St. Petersburg—it was winter time, and the room was much crowded. A gentleman accidentally broke a pane of glass, when a stream of cold air rushed in, and threw down the *moisture of the breaths of the company in snow flakes.*

"As snow in summer, and as rain in harvest; so honour is not seemly for a fool."—PROVERBS XXVI.

It was stated by a party of Dutchmen, who wintered in Nova Zembla, in the Arctic ocean, that whenever they went into the open air, the moisture of their breath was *converted into snow flakes* every time they breathed.

And it is stated by those who have travelled among the Esquimaux, who live in snow huts, that when the cold air rushes into their dwellings, *snow flakes are formed.*

401. *Why should we not throw salt upon the snow before our doors in winter?*

Because the salt, having a chemical affinity for water, dissolves the snow, and produces a kind of brine. But the temperature of this brine is considerably lower than that of ice or snow; and it produces a very unpleasant and chilling effect upon the feet of those who tread upon it.

402. *Why does the sunshine frequently feel very hot when snow is on the ground?*

Because the moisture of the atmosphere having been thrown down in the form of snow, and the cold of night continuing to precipitate the vapour of the air in the form of hoar frosts, there arrives a period when the air is free from such an amount of vapour as could form mist or cloud to *intercept the sun's rays.* There then prevails a clear dry atmosphere, through which the solar heat strikes with considerable power.

The white surface of the snow also reflects the sun's rays. Therefore in clear snowy weather the air is frequently felt to be unusually hot.

403. *When there has been a slight fall of snow, followed by a northerly gale, why does the snow disappear without melting?*

Because snow, though it is congealed, *evaporates*, and in the case mentioned the whole of the snow passes away by evaporation.

404. *Why may snow be formed into hard snow-balls?*

Because the numerous fine *points of crystals*, when the snow has been gradually formed, intersect each other and bind the snow into a compact mass.

405. *Why will not the snow bind into snow-balls sometimes?*

Because it has been congealed rapidly, and the crystal points, instead of being well developed, are *imperfectly formed*, like those of hail.

406. *What is sleet?*

Sleet is only a modification of snow, which, in falling, has met with a *warmer current of air* than that in which it congealed. It therefore partially melts and forms a kind of *wet snow*.

407. *What is hail?*

Hail is also the *frozen moisture of the clouds*. It is probably formed by *rain drops*, in their descent to the earth, meeting with an *exceedingly cold current of air*, by which they become *suddenly frozen into hard masses*, not having time to crystallize.

408. *Why does water in freezing expand?*

Because, when water freezes, its particles become arranged in crystalline forms. These crystals cross and intersect each other, and cause numerous interstices. Therefore, though the water parts with much of its caloric becoming ice, the arrangement of its atoms into crystals causes it to become of greater bulk. It also contains numerous air bubbles.

409. *Why does ice float upon water?*

Because, as it expands in freezing, it becomes *specifically lighter* than water.

410. *Why does the surface of the water being frozen prevent the frost extending to the depths of the water?*

Because ice, like snow, is a *bad conductor of heat;* the coating of ice therefore tends to keep the subjacent water warm.

411. *Why is ice generally full of air bubbles?*

Because, in freezing, some of the atmospheric air which is held in solution in the water *becomes imprisoned*.

" The waters are laid as with a stone, and the face of the deep is frozen."—JOB XXXVIII.

412. *Why is the ice which is found in agitated waters free from bubbles?*

Because, during the agitation of the water, the air *finds an opportunity of escaping* as the process of congelation advances.

413. *Why is ice considered to be the simplest state of water?*

Because in water there exists a considerable amount of *caloric* which is given out when water freezes; and air and some extraneous matters are expelled when water congeals.

414. *Why do vessels and pipes in which water is confined burst when the water freezes?*

Because the force exerted in the freezing of water is great and *irresistible.* The expansion of water in becoming ice is one of Nature's grandest laws, which she enforces with the greatest majesty. In the mountains of Switzerland, and in other countries, the rocks are frequently rent by the force of ice; and in severe winters the noise of the splitting of the crags is sometimes as loud as thunder.

415. *Why are the poles of the earth wrapped in ice?*

Because the sun's rays fall *obliquely* upon the polar extremities of the earth, by which the *intensity of their heat* is greatly diminished; and also because of the great length of the winter nights, which prevails throughout one-half of the year.

See Radiation of Light and Heat.

416. *Does ice exist in great density at the poles?*

Yes; its density at the polar regions is very great. Even in less severe latitudes the ice is so hard that *the tools used in cutting it* have frequently to be repaired.

417. Bishop Watson, in his *Chemical Essays*, gives the history of the marriage of Prince Gallitzi, of Russia, in 1739. The Russians built, for the celebration of the marriage, a palace in which ice served the purposes of stone. It consisted of two magnificent apartments; and even many of the articles of furniture were made of ice. Some cannon, of ice, were fired more than once without bursting.

"And God made two great lights; the greater light to rule the day, and the lesser light to rule the night: he made the stars also."—GEN. I.

418. *Why does frost benefit fallow soils?*

Because it expands the clods, and causes them to *break* and *crumble* into fine dust, exposing their matter to the air, enriching them by the absorption of gaseous matters.

419. *Is ice the only instance of water existing in a state of solidity?*

No; water becomes still more solid *in combination* with *lime* and other earths.

420. *Why does water become more solid in these combinations?*

Because it parts with more of its *caloric* than in the process of freezing, and is therefore the further removed from its natural state of fluidity.

CHAPTER XVIII.

421. *What is dew?*

Dew is *condensed vapour.*

422. *What causes the condensation of the vapour which forms dew?*

When the sun sets, the earth, and the various objects upon its surface, *begin to cool.* They throw off, *by radiation,* some of the heat which they received during the day, and, becoming *cooler* than the surrounding air, they *condense the vapour* of that portion of the air which comes in contact with them.

423. *Why do some bodies become wetter with dew than others?*

Because substances *differ in their powers of radiation.* Those which radiate the least heat remain comparatively *warm,* and do not favour the formation of dew upon their surfaces; while those which radiate most heat become *cold,* and cause the moisture of the air to settle upon them.

424. *Why is there more dew formed during some nights than on others?*

Because of the *variation in temperature,* and the manner in which those variations affect the relative heat of the earth, and the air in which the vapour is suspended.

425. *Is the formation of dew the cause, or the effect, of cold?*

It is the *effect* of cold. The coldness of the bodies with which the air comes in contact causes the vapour to condense, and form dew-drops.

426. *Does the formation of dew tend to cool or warm the air?*

It tends to warm the air, by *lessening the quantity of watery vapour* floating in it, and by giving to the air the *latent heat* which the vapour contained.

427. *Upon what substances does dew form most freely?*

Upon woolly, rough, and fibrous substances, such as the *wool* of animals, the *leaves* of plants, *blades* of grass, &c.

428. *Upon what substances does dew form least freely?*

Upon smooth surfaces, such as those of *metal, stones*, and *glass*

429. *Why does dew form upon rough surfaces more freely than upon smooth ones?*

Because rough surfaces *radiate heat freely*, and cause their substances to become cooled; while smooth surfaces *radiate less heat*, and, therefore, their temperature does not fall in the same degree.

430. *In what way does the radiation of heat from the earth by night illustrate the Divine wisdom?*

If the heat received by day were not radiated back by night, the earth, acquiring an addition of heat every day, *would soon become so hot* as to be unendurable by the vegetable and animal kingdoms now existing upon it.

431. *Why does the radiation of heat from rough surfaces exhibit the Divine wisdom?*

Because it causes the deposition of dew upon grass, plants, and

"Thou visitest the earth, and waterest it: thou greatly enrichest it with the river of God, which is full of water: thou preparest them corn, when thou hast so provided for it."—PSALM XLV.

trees; but upon barren rocks, gravel walks, and bare and hard soils, *where dew would be of no benefit*, it does not form.

432. *Why does little dew form at sea?*

Because of the low radiating power of the smooth surface of water; and also of the tendency of the water that has become cooled to sink, and allow warmer portions to rise, by which the temperature of the surface of the sea *is seldom greatly or suddenly reduced.*

433. *Why is little dew formed on windy nights?*

Because the air, being kept in rapid motion, *does not remain long enough in contact* with cooling surfaces to deposit its vapour.

434. *Why does a gentle breeze favour the formation of dew?*

Because, while it allows each stratum of air to remain in contact with the cold surface of the earth long enough to deposit its moisture, it *brings fresh strata* to the earth, by the movements of the air; and as these fresh strata bring *additional moisture*, there results an increased formation of dew.

435. *Why is a very calm night less favourable to the formation of dew than when a gentle breeze prevails?*

Because, through the stillness of the air, *the same portions of the atmosphere remain in contact* with the earth's surface; consequently, there is not so much vapour presented to the cooling bodies as when the air is in gentle motion.

436. *Why are clear and starlight nights favourable to the formation of dew?*

Because the radiation of heat from the surface of the earth is *not checked by the intervention of the clouds.*

437. *Why does the intervention of the clouds check the radiation of heat from the earth?*

Because the clouds *act as secondary radiators;* they receive the radiated heat of the earth, and *return it*, by which the degree of cold that would otherwise occur upon the surface of the earth is considerably modified.

" The Lord is my shepherd, I shall not want. He maketh me to lie down in green pastures."—PSALM XXXIII.

ILLUSTRATION OF THE FORMATION OF DEW.

438. If plates of glass be laid over grass-beds, as in the engraving, no dew will be deposited on the grass underneath the glass plates, although all around the grass will be completely wetted. The explanation is that the glasses, being radiators of heat, act in the same manner as the clouds, returning the heat to the bodies underneath them, and preventing the formation of dew thereon. It has been observed that sheep that have lain on the grass, during the formation of dew, have their backs completely saturated with it, but that underneath the line where their bodies turn to the earth, their coats are dry. In the same manner glass globes suspended in the air, on dew-forming nights, will be found loaded with globules of dew upon the top, but there will be no appearance of moisture underneath.

CHAPTER XIX.

439. *Why is there little or no dew formed when the clouds are dense and low?*

Because then the heat radiated by the clouds to the earth is considerable, and *prevents the cooling* of the surfaces upon which dew usually forms.

440. *Why is there an increase of dew when the clouds are very high in the atmosphere?*

Because the radiating effect of the clouds *diminishes with their distance* from the earth; and when they are far away, it is comparatively feeble.

441. *Why are islands, and places close to the sea, frequently warmer in winter than places removed far from the sea?*

Because, being subject to cloudy skies, formed by the evaporation

' Behold, I will put a fleece of wool on the floor; and if the dew be on the fleece only, and it be dry upon all the earth beside, then shall I know that thou wilt save Israel.

of the sea, the clouds *act as a screen*, and prevent the loss of the heat of the ground by radiation.

442. *Why are fogs unfavourable to the formation of dew?*

Because they *act in the same manner as clouds*, preventing the escape of radiant heat, and preserving the temperature of the surface of the earth.

443. *But why does it sometimes appear that much dew has been formed during a foggy night?*

Because the cold air of night *throws down the vapour of which the fog is composed*, and which, therefore, appears like dew. But real dew is that which is caused by the cooling of the earth; while the wetness caused by the precipitation of fogs by cold, is more allied to *rain* than to *dew*.

444. *Why is the gravel walk frequently free from dew while the grass is very wet?*

Because the gravel *radiates heat less freely than the grass*, and therefore it does not become so cold. Besides, the *conducting* power of bodies must be considered, as well as their *radiating* powers. A body might radiate heat, but if it were in a position to obtain fresh heat by conduction from other substances, it would still retain warmth.

445. *Why would a handful of gravel placed on a bed of grass become wetter than the gravel of the walk?*

Because the handful of gravel, *being cut off from the sources of heat by conduction*, and isolated from the greater mass of gravel, would radiate its heat, and fall to a lower degree of temperature.

446. *Why will more dew form on shavings of wood than on blocks of wood?*

Because the shavings, being thinner bodies, do not derive the same amount of heat by conduction. The surface of a block of wood would continually *draw off heat by conduction from its internal parts*. But the shavings, having no such sources to draw from, would become colder than the block.

"And it was so: for he rose up early on the morrow, and thrust the fleece together, and wringed the dew out of the fleece, a bowl full of water.

447. *Why do bodies that are elevated a little way in the air receive more dew than those that are in contact with the earth?*

Because such bodies as are in contact with the earth derive some degree of heat by *conduction* therefrom; while those that are elevated in the air are less favourably situated.

448. *Why is the opinion abandoned that dew proceeds from moisture driven off from the ground?*

Because it is unreasonable to suppose that the earth would be *warm* enough to *drive off vapour*, and *cold* enough to *condense it* at the same time.

Because, also, *hoar frosts* frequently occur upon the surface of *hard frosts*, when the surface of the earth being completely frozen, and the air being cold, *little evaporation* could have taken place.

In certain states of the soil, some portion of moisture is given off by evaporation from the earth, and this moisture afterwards becomes converted into dew.

449. *Why do the leaves of trees often remain dry throughout the night, while those of grass and shrubs are covered with dew?*

Because the air near the upper parts of trees is *more agitated* than that nearer the ground. The air is also warmer, because its temperature is less affected by the fall in the temperature of the surface of the earth And, from the pliancy of the twigs, the leaves are nearly always in motion.

450. *Why are valleys and sheltered places liable to heavier dews than elevated lands?*

Because valleys are characterised by *greater moisture;* and the air is liable to *less disturbance* than in elevated places.

451. *Does the fluid which is transpired by the leaves of plants contribute to the formation of dew?*

The transpired fluid of plants *contributes to the humidity of the atmosphere,* out of which dew is formed; but such transpired fluid is only a tributary part of dew, and not the direct cause of it.

"And Gideon said unto God * * Let it not be dry only upon the fleece, and upon all the ground let there be dew.

CHAPTER XX.

452. *Why is dew seldom, and only slightly, produced in cities?*

Because houses, and all other objects, *radiate heat.* The cooling of the earth is, therefore, materially checked by the abundant radiation from elevated objects.

453. *Why is dew formed more plentifully in spring and autumn than in summer?*

Because at those seasons the nights are longer, and there are great differences between *the temperature of day and night.*

454. *Why is dew formed most abundantly on those clear nights which are followed by foggy or misty mornings?*

Because the atmosphere possessed *a high state of humidity,* as is evidenced by the existence of fog or mist.

455. *Why is more dew formed between midnight and sunrise, than between sunset and midnight?*

Because the cold of the earth and the atmosphere is greater towards the morning, in consequence of *the number of hours through which heat has been thrown off* by radiation.

456. *Why does dew form rapidly on a clear morning after a cloudy night?*

Because, during the night, the clouds have prevented the formation of dew; the air has, therefore, *retained all its moisture*; and, as soon as the clouds pass away, the earth cools rapidly, and dew is deposited abundantly.

457. *Why do bedroom windows that are concealed by shutters on the inside, become covered with an unusual degree of moisture?*

Because the shutters prevent the glass from being affected *by the radiation of heat from the walls,* and the various objects within the room. It is, therefore, exposed to the cooling influence of the outer air, and the moisture of the air within the room condenses upon the cold glass.

458. *Why do shutters on the outside of the window prevent the deposition of so large an amount of moisture?*

Because, as well as excluding the cold air, the shutters *prevent the radiation of heat by the glass*, and, therefore, the moisture of the air in the room is not condensed upon it in the same degree.

459. *Why do the mats employed by gardeners protect plants from the effects of cold?*

Because they prevent the radiation of heat, not only by the plants, *but by the walls or other objects* by which the plants may be surrounded.

Whatever intervenes between an object and the sky, *prevents the radiation of heat from that object;* or *counterbalances the effect* of radiation by directing the heat back again.

460. *Why are plants, &c., frequently covered with hoar frost?*

Because the dew deposited upon the plants, &c., *freezes* from the cold.

461. *Why does hoar frost indicate the approach of rain?*

Because it shows that the air is *pregnant with moisture*, and the temperature being low, favours the *condensation of vapour* into rain.

462. *Why does hoar frost present to the eye such beautiful crystalline forms?*

Because the atoms of water, when deprived of the degree of heat necessary to keep them in a fluid condition, *arrange themselves in groups*, giving rise to what are called *crystals.*

463. *Why do the particles of frozen water thus arrange themselves?*

The cause is not clearly understood. But it is conjectured that *magnetism* influences the disposal of the particles of bodies in all the processes of *crystallization*—"that each particle of a crystalline mass has certain points or poles which possess definite properties, and that cohesion takes place only along lines which have some relation to the attracting or repelling powers of those poles."

464. *Why are the crystals of hoar frost more interesting in their forms than those of water frozen in masses?*

Because the minute drops of dew being, as it were, suspended in the air, the atoms of the drops are *free to arrange themselves* in crystalline groups, without the interference of resistance from the weight of surrounding particles, or other causes.

465. *Why do the crystals of hoar frost resemble those of snow?*

Because they consist of similar atoms crystallized *under somewhat similar causes.* There is a slight distinguishing difference between the crystals of hoar frost and those of snow, which may be accounted for by the various degrees of temperature in the atmospheric region where snow is formed, or by the different effects of the *magnetism* of the earth and the *electricity* of the air.

466. *What are fogs?*

Fogs are *clouds* which form on, or descend to, the *earth's surface.*

467. *Why are they generally dark and stifling?*

Because the aqueous vapours of which they are composed *mingle with the smoke and exhalations* of towns and cities. They, therefore, do not consist of pure watery vapour.

468. *Why are fogs most frequent in spring and autumn?*

Because the temperature is most *variable* in those seasons. Vapours become suddenly condensed by change of temperature.

469. *Why are fogs sometimes called dry fogs?*

Because they occasionally assume a dull *opaque appearance,* and produce a very *depressing effect* upon those who breathe them. Their peculiar darkness and oppresive effect have been attributed to *electricity.* But the cause of dry fogs is not clearly understood.

"Dost thou know the balancings of the clouds, the wondrous works of him which is perfect in knowledge."—JOB XXXIII.

CHAPTER XXI.

470. *What are clouds?*

Clouds are bodies of *vapour* floating in the atmosphere.

471. *What are the sources of atmospheric vapours?*

The vapours, of which clouds are formed, arise from *the evaporation of water* at the earth's surface.

472. *Why does water placed in a saucer disappear after a time?*

Because it passes away by *evaporation.*

473. *Why does ink become dried up in ink-stands?*

Because its *watery particles* evaporate.

474. *Why do ponds become dried up in dry weather?*

Because the water passes away by evaporation. Ponds of considerable size and depth frequently become *dried up.*

475. *Why does wet linen become dry by exposure to air?*

Because the water *flies away* by evaporation.

476. *What becomes of the water evaporated from the saucer, the ink, the pond, &c.?*

It contributes to the *formation of clouds.*

477. *What other sources of evaporation are there?*

The chief source of evaporation is *the sea* especially along its borders, where it becomes heated by the earth; besides which *rivers* and *lakes,* the *steam* from manufactories and locomotives, and the water produced by the breathing of animals and the burning of fires, gas, &c., all contribute to the *formation of clouds.*

478. *Why does vapour ascend?*

Because, being highly rarefied by heat, it is *specifically lighter* than the air.

"With clouds he covereth the light, and commandeth it not to shine by the cloud that cometh betwixt."—JOB XXXVI.

479. *In what state does vapour exist in the clouds?*

By some persons it is supposed to form *watery vesicles,* or minute bubbles; by others it is believed that the diffusion of caloric between the particles of the vapour forms a kind of *aqueous gas.*

480. *Why does rain sometimes suddenly set in?*

Because a sudden change of temperature produced by the advance of *cold currents of wind,* or other causes, may deprive the watery vapour of its caloric, and cause the particles of the vapour to *unite with each other,* forming drops of water that fall in the form of *rain.*

481. *Why, when a decanter of cold water is brought into a room, does it soon become clouded with moisture?*

Because the cold decanter *attracts the caloric from the vapour;* as the caloric abandons the vapour, and passes into the decanter, the vapour condenses, and forms small *drops of water.*

In this illustration, the cold decanter represents a cold current of air acting upon the air, and producing clouds.

482. *Why do clouds present such varied appearances?*

Because they are affected by the degree of moisture in the air; by the changes of temperature; by the direction of the winds; by the heights at which they fly from the earth; by the electrical states of the air; and by their positions in relation to the light of the sun and moon.

483. *How many forms of clouds have been defined and named?*

There are three simple forms, and four compound forms.

1. Cirrus
2. Cumulus.
3. Stratus.
4. Cirro-cumulus.
5. Cirro-stratus.
6. Cumulo-stratus.
7. Nimbus, or Cumulo-cirro-stratus.

"Behold, he withholdeth the waters, and they dry up; also he sendeth them out, and they overturn the earth."—JOB XII.

484. *Why are some clouds called cirrus?*

The Latin word *cirrus* means a lock, or curl of hair, the long hair of boys. This name has, therefore, been given to those *thin curly clouds* that either in long curly streaks, or in occasional curly groups, appear in the atmosphere.

CIRRUS, OR CURL CLOUD.

485. *What are the peculiarities of cirrus clouds?*

They generally appear in *fine weather*, and are said to indicate the continuance of the same weather, or a change, according to the phenomena which they exhibit.

If at first they *appear high*, and then become *lower and denser*, they indicate the coming of *stormy weather*, especially if they repeat this indication two evenings or more in succession. The storm is said to come from the point opposite to the clouds.

If they spread across the sky in *long light streaks*, they indicate the coming of *wind* in the direction of the streaks.

If they resolve themselves into *long streaks*, which gradually spread into *sheets*, and then merge into the *cirro-stratus* form, then continued *wet weather* may be expected.

If they *fly high*, and retain their distinguishing form; if they do not rapidly *increase* in number, or *alter* in shape, they indicate the continuance of *fine weather*.

486. *Why are other clouds called cumulus?*

The Latin word *cumulus* means a heap, an accession, or an addition; it has, therefore, been applied to those clouds that *grow out of a small nucleus*, and enlarge until they acquire the appearance of vast bodies.

CUMULUS, OR PILE CLOUD.

487. *What are the peculiarities of cumulus clouds?*

They are formed in the lower atmosphere, and move along with the current that is next the earth.

A small spot at first appears, and is the nucleus on which they increase. The lower surface forms an irregular plane, but the upper parts rise into round fleecy heaps.

If their formation is gradual; if they increase and rise about noon, and disappear towards sunset, they indicate *fine weather*.

If they change frequently and suddenly, their fleecy parts breaking into thin streaks, and their general character changing to the compound forms, they *foretell rain*.

If they continue to increase, and rise about sunset, they indicate *storm and thunder* in the night.

488. *Why are other clouds called stratus?*

The Latin word *stratus* means strewed, scattered, laid long, flat, prostrate; it has, therefore, been applied to the lowest clouds, whose inferior surfaces commonly rest on the earth or water; including all those creeping mists which in calm evenings and mornings ascend in spreading sheets (like an inundation of water) from the bottom of valleys, and the surface of lakes, rivers, &c.

"Terrors are turned upon me: they pursue my soul as the wind; and my welfare passeth away as a cloud."—JOB XXX.

489. *What are the peculiarities of stratus clouds?*

If, when they appear in low places, they gradually put on the appearance of *cumulus*, and rise, and either evaporate or fly away, they foretell *fine weather*.

But if *cumulus* clouds merge into *stratus*, and descend, then *damp and cheerless weather* may be expected.

490. *Why are some clouds called cirro-cumulus?*

Because they are formed by a combination of the *cirrus* and the *cumulus*.

CIRRO-CUMULUS CLOUDS.

491. *What are the peculiarities of cirro-cumulus clouds?*

They form a very beautiful sky, frequently exhibiting numerous distinct beds of small clouds, floating at different altitudes.

In summer they foretell *heat* and *dry weather*, if they fly high, and in broken, well-defined masses. But if they gather low, and merge into the *cirro-stratus*, they foretell *change to wet*.

492. *Why are some clouds called cirro-stratus?*

Because they are formed by a combination of the *cirrus* and the *stratus*.

493. *What are the peculiarities of cirro-stratus clouds?*

They are generally first formed by the *cirrus* clouds reclining to a horizontal position, and drawing towards each other sideways. Their structure varies much, presenting at times the appearance of parallel bars, thickest in the middle or at one extremity, and growing thinner towards the edges. They are frequently intermingled with small particles, giving to the sky the appearance of grained wood.

CIRRO-STRATUS CLOUDS.

The *cirro-stratus* precedes wind and rain, or snow, the near or distant approach of which may be estimated by the greater or less abundance and permanence of the clouds.

Sometimes the *cirro-stratus* and the *cirro-cumulus* appear together in the sky, and even alternate with each other in the same cloud, when the rapid changes which ensue afford a curious spectacle, and an opinion of the weather which is likely to follow may be formed by observing *which form of cloud prevails at last.*

The cirro-stratus is the form of cloud which most frequently exhibits solar and lunar *haloes.* Hence the reason why *foul weather* is expected when haloes appear.

494. *Why are some clouds called cumulo-stratus?*

Because they are formed by a combination of the *cumulus* and the *stratus.*

495. *What are the peculiarities of cumulo-stratus clouds?*

They are frequently seen in the same sky with the *cirro-stratus,* the latter forming dull streaks running across the mountain-like pile of the *cumulo-stratus.*

"Canst thou lift up thy voice to the clouds, that abundance of waters may cover thee?"—JOB XXXVIII.

CUMULO-STRATUS, OR TWAIN CLOUD.

They generally form lofty and dense piles of cloud, presenting to view the imagery of a beautiful landscape. If they rise and become light, and fleecy, they foretell *fine weather;* but if they merge into the *cirro-stratus, unfavourable weather* may be expected. When they lower, and instead of rising high and appearing fleecy, assume the *nimbus* form, they denote the approach of *thunder-storms.*

496. *Why are some clouds called nimbus, or cumulo-cirro-stratus?*

The Latin term *nimbus* means a storm-cloud; a violent storm of rain, &c. These clouds are formed by combinations of the *cumulus, cirrus,* and *stratus.*

497. *What are the peculiarities of nimbus clouds?*

The characteristics of *nimbus* clouds can best be seen when they are at a distance, for then the combination of the *cumulo-cirro* forms, over a dense mass of *stratus,* can be distinctly witnessed. When they fly over head, the latter can only be recognised by occasional openings in the dense *stratus,* through which the lighter forms of the *cumulus* and *cirrus* may be seen.

When these clouds gather rapidly, they bring passing storms of *rain, hail,* or *snow.*

"When he made a decree for the rain, and a way for the lightning and the thunder."—JOB XXVIII.

NIMBUS, OR STORM CLOUD.

When they come with the wind, and form dense masses of black cloud, *lightning and hail* may be expected.

When their position is opposite to the sun, a *rainbow* generally appears, which looks all the more beautiful in contrast with the dense mass of cloud behind it.

CHAPTER XXII.

498. *Why do clouds sometimes appear to rise, fall, and revolve in the upper regions of the air?*

Because sometimes warm currents of air ascend, and when they reach the higher and colder regions their vapour condenses and falls. As it falls it again comes in contact with warmer air, is *again rarefied*, and ascends.

499. *Why does it sometimes seem to rain from the clouds, without drops of rain falling to the earth?*

Because when the temperature of the upper and lower atmospheres are disturbed by moving currents, rain or snow may be formed and precipitated from the upper atmosphere, but be *evaporated again* upon reaching a warmer stratum of air. Hence it may rain from the clouds, but the rain may not reach the earth.

500. *Why do clouds generally gather and obscure the sky before a considerable fall of rain?*

Because when the vapour of the air is small in amount, upon

reaching the higher regions of the atmosphere it produces the light *cirriform* clouds; but as the vapour augments, and its condensation advances in the upper regions, its weight, relative to that of the atmosphere, increases, and it subsides to a lower station, where, met by a denser atmosphere, it floats as a cloud of the *cirro-cumulus* or *cumulo-stratus* form.

501. *Why do clouds frequently gather and disperse several days in succession before a fall of rain?*

Because when the clouds obscure the sky, they arrest some of the heat of the sun's rays, and they also receive the heat that is radiated from the earth. Consequently, when they gather, they themselves produce the *cause of their dispersion*, and unless the air is fully saturated with moisture, the clouds may pass away and gather again several times before a fall of rain.

502. *Does water evaporate at all temperatures?*

Yes, *even ice and snow evaporate.* Millions of tons of water are evaporated from the surface of the earth every day. By this evaporation the earth is refreshed, temperatures are equalised, rivers are supplied, and the depth and extent of oceans are considerably influenced and regulated.

503. *Why do clouds of dust indicate the coming of rain?*

Because evaporation takes place much *more rapidly during the prevalence of wind* than when the air is still. Clouds of dust not only indicate the dryness of the earth, through the absence of rain, but they also show that winds are prevailing, and that evaporation by those winds is actively going on; therefore the air will the sooner be saturated with moisture.

504. *Why does not evaporation proceed so rapidly in a calm as in a breeze?*

Because when air is still the same portions of it remain in contact with the watery surfaces; but when breezes prevail fresh and dry volumes of air are repeatedly *brought in contact with the evaporating surfaces.* It is for this reason that wet linen dries more rapidly in a wind than in calm air.

"Who giveth rain upon the earth, and sendeth waters upon the fields."—JOB v.

505. *Why do the different phases of the moon affect the quantity of rain?*

Because the influence of *the moon's attraction* alters the density of our atmosphere. From observations made during nineteen years, it has been found that the average number of rainy days for each of the moon's phases is as follows: new moon, 77; first quarter, 82; full moon, 79; last quarter, 60; moon in perigee, 93; moon in apogee, 78.

Perigee—the point at which the moon is least distant from the earth. *Apogee*—her most distant point from the earth.

CHAPTER XXIII.

506. *Why do some substances feel warm and others cool?*

Those that feel warm are generally of a *higher temperature* than our bodies; and those that feel cool are of a *lower temperature.*

But the *sensations of heat and cold* are influenced by the *conducting powers* of substances. Iron, though of the same temperature as wool, *feels differently to the touch,* because its conducting power is much greater than that of wool.

507. *Why, when we bathe in the sea, does the water appear to be colder than the air?*

Because water abstracts heat from the body *more rapidly than air.*

508. *Why does a still atmosphere in summer feel warm?*

Because those particles of air which are in contact with the body *acquire the temperature thereof,* and through the stillness of the air they bear away the heat slowly.

509. *But why, if a wind arises, does it feel cool?*

Because fresh particles of air are continually brought to the surface of the body, and are no sooner warmed than *they are displaced by cool particles,* and thus *a feeling of coolness* is produced.

510. *Why do sheets feel colder than blankets?*

Because they are *better conductors of heat.*

"Say unto wisdom, Thou art my sister; and call understanding thy kinswoman."—PROVERBS VII.

511. *But why, when the sheets acquire warmth from the body, do they feel warmer than the blankets?*

Because, being better conductors of heat, and the heat being accumulated by them, they *impart a greater sensation of warmth* when they come in contact with the body.

512. *Why does flannel, worn next to the skin, form a warm covering in cool climates, and a cool covering in warm climates?*

Because, being a slow conductor, it *keeps in* the warmth of the body when the external air is cold; but it *keeps out* the external heat when the air is hotter than the body.

513. *Why does the use of a fan produce a sensation of coolness?*

Because it *keeps the air in motion* over the surface of the face, and each wave of air bears away some portion of the heat of the body.

514. *But if the air were hotter than the body, what would then be the effect of the fan?*

Then fanning the face would produce *a sensation of increased heat.*

515. *Why does glass, though an indifferent conductor of heat, feel cold to the touch?*

Because it is so dense that it *presents a great many particles* to the touch, and each particle takes up some portion of heat. But the heat will not pass away rapidly, glass being a poor conductor. Hence glass, which at first feels cool, soon ceases to impart that sensation, but *creates a feeling of warmth.*

516. *Why are wood or ivory handles put upon metal teapots?*

Because wood and ivory are poor conductors of heat; they therefore *prevent the heat* from being transmitted by the metal to the hand.

"It is a good thing to give thanks unto the Lord, and to sing praises unto thy name, O Most High."—PSALM XCII.

517. *Why have porcelain teapots handles of the same material?*

Because porcelain is an indifferent conductor of heat, and does not create inconvenience by *conveying heat rapidly to the hand.*

518. *Why, when metal teapots have metal handles, is there a ring of wood or ivory inserted where the handle is joined to the teapot?*

Because the intervention of the non-conductor *interferes with the transmission of heat* from the body of the teapot to the handle.

519. *Why are cloth and paper employed as kettle-holders?*

Because the metal handle of a tea-kettle cannot be touched by the hand without inconvenience, *unless some such non-conducting substance is interposed.*

520. *Why may a man stand in an oven at a very high temperature?*

Because the *air* surrounding him is a *poor conductor;* and if he stands upon a *matting* of *wool,* or *straw,* the heat will be conducted to him but slowly. If, however, under such circumstances, he were to *touch a metal body,* he would be seriously burnt.

521. *Why will a bright silver teapot make better tea than an earthenware one?*

Because bright surfaces *radiate,* or throw off heat much slower than black and dull ones, consequently the tea is kept hotter.

522. *Why will a saucepan which is black on the part exposed to the fire, but bright on the portion in contact with the air, boil quickest?*

Because the *black part* acts as a good conductor of heat from the fire; and the *bright part,* being a bad radiator, will not diffuse that heat in the surrounding air.

"Dost thou know the balancings of the clouds, the wondrous works of him which is perfect in knowledge."—JOB XXXVII.

523. *Why should woollen substances be wrapped around anything that is to be kept hot?*

Because it is a bad conductor, and will *retain the heat.*

524. *Why is ice preserved from melting by wrapping it in flannel?*

Because the flannel will not *conduct the external heat* to the ice.

525. *Why does sugar take a long time to dissolve in the bottom of a cup of tea?*

Because the sugar, as it dissolves, *renders the tea heavier*, and impairs the solvent power of the water by which the sugar is surrounded.

526. *Why does stirring the tea with a spoon cause the sugar to dissolve?*

Because it *directs fresh particles* of water to the sugar and diffuses the atoms of sugar between the atoms of water.

CHAPTER XXIV.

527. *Why does the bright tin screen which is placed behind meat while roasting, increase the effect of the heat?*

Because it *prevents the heat from escaping* into the kitchen, by reflecting it back towards the joint and the fire.

528. *Does an inverted cup, in a pie, prevent the juice from boiling over?*

It does *not*. When the cup is first placed in the pie it is full of cold air, which *expands as it becomes hot,* and tends rather to force the liquid out of the pie, instead of keeping it in.

529. *But why is the cup generally full of juice when the pie is set upon the table?*

Because, when the pie is taken out of the oven, the rarefied air that remains within the cup *cools and condenses.* It then occupies much less space, as a portion of the air which formerly filled the cup has been driven away; and now the pressure of the external air forces the juice into the cup as the air within it condenses.

530. *Why, then, are pies into which inverted cups have been put more juicy than those from which cups have been omitted?*

Because, while the pie is cooling, a great deal of its moisture *evaporates.* But, as much of the juice is forced into the cup, it cannot evaporate; and, therefore, *the juice is preserved.*

531. *Why do broths and syrups become stronger by boiling?*

Because heat *drives off a portion of the water* in the form of steam; but the *solid matter* still remains, and gives increased strength to the diminished quantity of fluid.

532. *Why do spiritous mixtures become weaker by boiling?*

Because alcohol, which is the spirituous part, is *very volatile* and evaporates before the water boils, leaving the water behind.

533. *Why is it difficult to heat water from the top?*

Because heat is transmitted through water by the *movement of its particles*. And as those particles which are heated are lighter than those that are cold, they *remain at the top* (when heat is applied there), and do not convey heat to the other parts of the water.

534. *Why does boiling water bubble?*

Because the lower parts of the water, being considerably *expanded by heat,* rise rapidly to the surface, while the cooler portions descend. The boiling of water consists of a rapid movement of its particles from the bottom to the surface, and down again, as the hotter particles displace those that have become cooled.

"He casteth forth his ice like morsels: who can stand before his word?"—
PSALM CXLVII.

535. *Why should ice, which is used for cooling water, &c., be applied at the top of the water?*

Because the cold particles of the water, being *heavier than the warmer portions,* would descend, and the whole of the water would therefore be presented to the cooling effects of the ice.

536. *Why does ice float?*

Because it is lighter than water, in consequence of *crystallization* having expanded its particles.

537. *Why would great evils arise if ice were heavier than water?*

Because as each successive layer of ice formed, it would sink *to the bottom,* until the whole depth of the ocean and the lakes became frozen. The fishes of the sea would be destroyed; the sun would scarcely have power to penetrate these vast masses of ice; and the thaw would be attended with dreadful effects, even more serious than those which are now encountered in the northern seas.

538. *Why does ice, when lakes and ponds are frozen, tend to keep the water underneath it warm?*

Because it forms a complete covering, by which the *surface of the water is sealed,* so that it cannot *part with the heat* which it derives from the *earth.*

539. *Then why would water be cooled in small quantities by ice applied at its surface, while it is said that the frozen surface of lakes, &c., tends to keep the water warm?*

Because, in the former instance, there is no sources of warmth *acting upon the lower part of the water;* nor would its surface be *completely covered,* as in the frosts of nature; nor would the *surrounding air* be cold enough to freeze the water.

540. *Why, when we descend into a cave in summer, do we feel cold?*

Because the air of the cave, being sheltered from the rays of the sun, and the radiation of heat by heated surfaces, is *colder* than the *external air.*

"From the place of his habitation he looketh upon all the inhabitants of the earth."—PSALM XXXIII.

541. *Why, if we descend into the same cave in winter does the atmosphere appear to be warm?*

Because the external air is then *colder* than the air of the cave; and therefore, though the air of the cave may be at the *same temperature* as when it felt cold in the summer, it imparts to us *a sensation of warmth.*

CHAPTER XXV.

542. *What causes the frost-crystals upon windows?*

The *vapour* in the air within the room coming in contact with the glass of the windows, which is cooled to the freezing point by the temperature of the external air, *freezes,* and in congealing its minute particles arrange themselves into *crystalline forms.*

543. *Why does this more frequently occur in bed-rooms than in other apartments?*

Because the *breath of persons sleeping* in the confined air of bed-rooms imparts to the air a considerable amount of *moisture;* and also because the temperature of the external air is generally *lower by night* than by day.

544. *Why do the crystals disappear as the day advances?*

Because the warmth of the sun's rays *gives back to the ice-crystals* the degree of *heat* necessary to liquefy them.

545. *Why does the return of heat to the ice-crystals cause them to dissolve?*

Because the heat *alters the condition of the atoms* or particles composing the crystals, and becoming *latent,* or hidden among those particles, produces a *liquid* where a *solid* previously existed.

546. *Why does the melted ice run down the windows in streams of water?*

Because the water obeys the law of *gravitation,* which draws all bodies towards the centre of the earth.

547 *Why does some portion of the water adhere to the glass?*

Because it is held to the glass by *attraction*, which, under some circumstances, causes bodies even of dissimilar natures to adhere together.

548. *Why does the water form into globules when it descends upon the window-frame?*

Because the particles of the water *cohere*, or hold together, while the wood and the putty of the frame *repel* them; and until that property of *repulsion* is overcome between the water and the wood by contact, the water stands in globular drops.

549. *Why do the drops flatten and disappear after a time?*

Because, when the *repulsion* between the water and the wood has been overcome by contact, *attraction* takes place; the wood which previously *repelled* the water now *attracts* it. The water, therefore, spreads itself along the surface of the wood, and is partly *absorbed* by it, while the other part of the water, in contact with the air, is *evaporated*, and passes away.

550. *Why does a small drop of water run slowly down the glass?*

Because the *attraction* between the water and the glass is strong enough to almost overcome the *attraction of gravitation*, which draws so small a body of water to the earth.

551. *But why, when two drops unite, do they run down the pane more rapidly?*

Because then the *attraction of gravitation* acting upon the larger body becomes relatively greater than the attraction acting between the water and the glass. The balance of force in favour of gravitation *is increased*, and the enlarged drop runs more rapidly down the pane.

"Be not as the horse, or as the mule, which have no understanding: whose mouth must be held with the bit and bridle."—PSALM XXXII.

552. *Why, upon stepping out of bed, does the marble hearth feel cold to the feet?*

Because it is *a good conductor of heat,* and by *drawing off* heat from the foot produces therein a *sensation of cold.*

553. *Why does the carpet feel warmer than the marble hearth?*

Because, being *a bad conductor of heat,* it does not bear away the warmth of the foot so rapidly as the hearth, and therefore it produces relatively *a sensation of warmth.*

554. *If the marble and the carpet were both warmer than the body, what sensation would they then produce?*

The marble would then appear to the touch to be warmer than the carpet; because, being a good conductor, it would deliver heat to the body *more freely* than the carpet.

555. *Why do sheets feel cold while blankets feel warm?*

Because linen is *a better conductor* of heat than *flannel,* and by bearing away the heat of the body, it imparts *a sensation of coldness.*

556. *Why does a linen shirt feel colder than a cotton one?*

Because the *fine and close fabric* of the linen bears away the heat of the body more rapidly than the *looser texture* of the cotton.

557. *Why does water feel cold to the hand?*

Because it is a *better conductor* of heat *than air,* and draws off the warmth of the hand more rapidly.

558. *Is water a good conductor of heat?*

It is *not* considered a good conductor, because heat will not readily pass through it by the ordinary course of conduction. But as the particles of water become warm, *they move,* and cold particles take their place. Hence water, *by the movement of its particles,* bears away as much heat as if it were a good conductor.

"I will praise thee, O Lord, with my whole heart; I will show forth thy marvellous works."—PSALM IX.

559. *Why is air not so good a conductor of heat as water?*

Because its particles, within a given space, are *less numerous* than those of water. It does not, therefore, present to the same surface *so many carriers of heat* as water does. But it acts in the same manner as water in the removal of heat.

560. *What is the relative temperature of various articles in a room?*

Assuming that they all stand in the same relation to the sun, or the other *sources of heat*, they are *all of the same temperature*, though some may feel much warmer than others.

CHAPTER XXVI.

561. *Why are bed-curtains injurious to health?*

Because they *confine around the body* of the person within them the air which *has been deprived of its oxygen*; and they *prevent the fresh air* of the room from approaching the person who requires to breathe it.

562. *What kind of air is imprisoned within the curtains?*

Air charged with an unhealthy proportion *of carbonic acid gas*, and an excess of *nitrogen*, which has been thrown off from the lungs of the breather.

563. *Why should beds with curtains be open at the top?*

Because then the warm and foul air would *rise and pass away*, and fresh air would enter to supply the demands of life.

564. *Why do we pause in breathing after each expiration of the breath?*

The pause allows time for the corrupt air, which is thrown out, to *rise and pass away;* and also for the oxygen then in the lungs to unite chemically with the carbon of the blood, and form *carbonic acid gas.*

"I will both lay me down in peace and sleep: for Thou, Lord, only makest me dwell in safety."—PSALM IV.

565. *Why is air, which has been once breathed, unfit to be breathed again?*

Because it has parted with a large proportion of its oxygen, which it has *communicated to the blood*, and has *taken up carbonic acid gas*, which it received from the blood.

566. *How is impure air purified?*

Chiefly through the agency of the *vegetable kingdom*. The leaves of plants may be said to breathe air in a manner somewhat similar to the lungs of animals; but plants *absorb carbonic acid gas*, and *set oxygen free;* while animals *absorb oxygen*, and *set carbonic acid gas free.* Animals and vegetables, therefore, reciprocate each others' wants.

567. *How may we prove that plants give back oxygen to the air?*

If we invert a tumbler, and place a lighted taper under it, the taper will burn *until the oxygen is consumed*, when it will go out. The glass will then contain air, almost destitute of oxygen, and charged with *excess of carbonic acid gas.* If we then place the tumbler carefully over a sprig of a growing plant, in such a way that the deteriorated air cannot escape, nor intermix with the external air, the plant will absorb the *carbonic acid gas*, and restore *oxygen* to the air in the glass. After a few days the taper will be found to burn in the glass as freely as it did before the experiment.

568. *Why does fresh air impart a healthy glow to the cheeks?*

Because it *purifies the blood* by the free introduction of oxygen, which gives a red colour to the vital fluid.

569. *Why do the inhabitants of crowded cities generally look pale and emaciated?*

Because they breathe impure air; and in many respects their mode of life is *incompatible with a state of healthy existence.*

"O Lord, how manifold are thy works, in wisdom hast thou made them all: the earth is full of thy riches."—PSALM CIV.

570. *What is animal heat?*

Animal heat is that which *gives warmth to animal bodies* through the agency of their *internal* organisations.

571. *Why is heat generated within animal bodies?*

Because the mixture of the oxygen of the air with the carbon of the blood *produces a kind of slow combustion*, which is attended by the development of heat.

572. *Whence is the carbon of the blood derived?*

It is derived from *the food we eat.* Food may, therefore, be regarded as the *fuel* which feeds the fire, or the lamp of life.

573. *Why is the temperature of the human body the same in cold as warm latitudes?*

Because that degree of temperature is essential to the maintenance of the vital functions. It has, therefore, been ordained by the Creator that the denser air of the polar regions shall convey *a larger quantity of oxygen* to the lungs of the inhabitants of those cold regions; and that the highly carbonised food, such as train oil and blubber, which the climate enables them to eat, shall supply *an increased quantity of carbon* to sustain the more active internal combustion necessary to support the temperature of their bodies.

574. *Why does exposure to cold produce hunger?*

Because the heat of the body, being conveyed away, there is a *greater demand made upon the fires of the body* to keep up its warmth. And as food constitutes the fuel, it is demanded in increased quantity.

575. *Is animal heat really similar to the heat of the fire?*

Yes. All heat results from the presence of *caloric*, though the degrees of heat depend upon the various circumstances and proportions in which caloric appears. There is as much heat in the body of man as would, if it were compressed into a smaller space, cause a *bar of iron*, of considerable size, to become *red-hot*.

"Oh that men would praise the Lord for his goodness, and for his wonderful works to the children of men."—PSALM CVII.

576. *Why does exercise increase the heat of the body?*

Because it *accelerates the circulation of the blood* for the purpose of sustaining the bodily effort required. And, as the blood circulates, more carbon is brought by the blood to the lungs, and the quicker breathing supplies an increased amount of oxygen to combine with the carbon, and therefore greater heat is evolved?

577. *Why, if we quicken our breathing while sitting still, do we not increase the heat of our bodies?*

Because, although we force air into the lungs, we *do not increase the circulation of the blood.* Therefore, the oxygen cannot combine with the carbon in any increased degree.

578. *Why do poor and ill-fed people generally dislike fresh air?*

Because the oxygen of the fresh air demands from their bodies a supply of fuel to the lungs which *their scanty food cannot yield sufficiently.* They, therefore, prefer ill-ventilated hovels to the purer air.

CHAPTER XXVII.

579. *Why is spring water frequently hard?*

Because, by passing through the fissures of the earth, it comes in contact with mineral substances, and takes up in solution some of their particles. The existence of the minute particles of earthy matters among the atoms of water have a tendency to *destroy its fluidity* and softness.

580. *Why is rain water soft?*

Because, being distilled from the clouds, it is free from earthy matters, and its atoms *possess their natural fluidity.*

581. *Why does the soap used in washing the hands float in flakes upon the top of hard water?*

Because the lime, held in solution by the water, *unites with the tallow*, and forms an insoluble compound. The sulphate of lime *neutralises the soda* by combining with it, and therefore the cleansing power of the soap is impaired.

582. *Why does carbonate of soda, or carbonate of potash, render hard water soft?*

Because it *combines with the lime* held in solution in the water, and *precipitates* it. The soda, or potash, to be fully effective, should be dissolved in the water twenty-four hours previous to using it. The matters that harden the water would then sink to the bottom.

583. *Why does a soap-bubble display prismatic colours?*

Because the rays of light that fall upon the bubble are refracted by it: and, as the thickness of the film changes, the degree of refraction varies, and thus the "play" of beautiful colours is produced.

584. The refraction of light, and the production of prismatic colours, surround us with most interesting phenomena. The laundress, whose active labours raise over the wash-tub a soapy froth, performs inadvertently one of the most delicate operations of chemistry—the chemistry of the imponderable agents—and the result of her manipulations manifests itself in the delicate colours that dance like a fairy light over the glassy films that follow the motion of her arms. The laughing child, throwing a bubble from the bowl of a tobacco-pipe into the air, performs the same experiment, and produces a result such as that which filled the philosophic Newton with unbounded joy. The foam of the sea-shore, the plumage of birds, the various films that float upon the surface of waters, the delicate tints of flowers, and the rich hues of luscious fruits, all combine to remind us, that every ray of light comes like an angelic artist sent from heaven, bearing upon his palette the most celestial tints, with which to beautify the earth, and show the illimitable glory of God.

585. *Why can bubbles be best formed by the aid of soap?*

Because the soap, being diffused among the atoms of water, establishes between them a degree of *tenacity*, or *cohesive power*, which does not exist in water alone in sufficient force.

586. *Why do soap-bubbles ascend?*

Because, being filled with the warm air expired from the lungs, they are *lighter* than the colder external air. They are, therefore, *pressed upwards* by the denser air which surrounds them.

"He hath made his wonderful works to be remembered: the Lord is gracious and full of compassion."—PSALM CXI.

587. *Why do soap bubbles descend after a time?*

Because the air within them becomes *condensed by cooling;* and as it contains a considerable proportion of *carbonic acid gas*, it is, together with the film of the bubble, heavier than the surrounding air.

588. *Why does steam issue from the spout of a kettle?*

Because the heat of the fire *passes into the water*, and tends to *separate the particles* of which the water is composed. The heated portions of the fluid rise to the top, and when a certain degree of heat has entered the water, the cohesive force which held the particles together is so far overcome that some of the atoms fly away in the form of *vapour*.

589. *Why does the water in the kettle become hot?*

Because it *absorbs caloric* from the fire.

590. *What becomes of the steam that flies away from the kettle?*

It diffuses itself through the air of the room, until, meeting with a cold surface, *it condenses*, and returns to its original state of *water*. Or it passes into the outer air, and assists in the *formation of clouds*.

591. *If a tumbler be held near the spout of the kettle, why will it become covered with moisture?*

Because the cold glass *condenses the steam*, which forms minute drops of water that attach themselves to the glass.

592. *Why, when cold glasses and decanters are introduced into warm rooms, do they frequently become dulled by moisture?*

Because the vapour held in suspension in the air becomes cooled, and *condenses* upon them.

593. *Why are southerly winds usually rainy?*

Because they reach us from warm regions, and as they *come across the sea*, they imbibe a large amount of moisture; then, when they *meet the colder* air of our latitude, the moisture condenses, and falls as rain.

"He causeth the vapours to ascend from the ends of the earth; he maketh lightnings with rain: he bringeth the wind out of his treasuries."—JEREMIAH X.

594. *Why does the vapour of hot water rise up on the air?*

Because it is so attenuated by heat, that is *specifically lighter than the air*, and therefore rises and floats upon it.

595. We frequently hear of the singing of a kettle; but do we fully comprehend the music of its song? Through how many centuries had that grim old bard chaunted his lay of progress, unheeded by the faithless groups that sat around, until the Marquis of Worcester, in the year 1663, published his account of "a fire water work," which he had discovered, and told how he had burst an old cannon by filling it with water, and "screwing up the broken end, as also the touch-hole;" and how, by "making a constant fire under it, within 24 hours it burst and made a great crack." Doubtless the kettle had tuned its song many times to the Marquis, until one day it sang with greater effect, and called attention to those portentous notes that it had so frequently uttered. *Then* the song of the kettle was one of complaint; *now* it is a hymn of triumph. Less than two centuries have passed since the discovery of the "fire water-work" was announced, and now the kettle in various improved forms is working wonders in most parts of the world. The offspring of the old kettle contend with the mighty ocean and the tempestuous wind; they fly with bird-like speed over the face of the earth, and in practical effect expand time and shorten space; they plough, they dig, they weld, they weave, they clothe and feed the labourer, they enrich the capitalist, they protect the state. These glorious results were brought to pass through a Marquis listening to a tea kettle and interpreting its song! Thus it is shown that there is no voice in nature too humble to invite the ear; nor any human being so high but he may gather wisdom from nature's simplest lessons.

CHAPTER XXVIII.

596. *Why do air and smoke ascend the chimney?*

Because the air, as it sweeps through and over the fire, becomes rarefied, or expanded by heat; it is therefore *specifically lighter* than the surrounding air, which, pressing towards it, forces it up the chimney in a continuous stream.

597. *Why does air, when there is a fire in a room, rush in through the crevices of the doors and windows?*

Because, as the rarefied air ascends through the chimney, it disturbs the balance of density *between the air in the room and the outer air*. The dense and cold air, therefore, rushes in and displaces the rarefied and warm air.

"For my days are consumed like smoke, and my bones are burned as an hearth."—PSALM CII.

598. *Why do light and rarefied bodies ascend?*

They ascend by the *pressure of surrounding bodies* acting upon them, and forcing them upwards.

It is a mistake to suppose that any bodies have of themselves a tendency to fly upward.

Even hydrogen gas, the lightest of all known bodies, has weight, and would fall to the earth, if not surrounded by the denser air.

599. *Why are chimney pots set upon chimneys?*

Because, by contracting the aperture of the chimneys, and allowing only space for the ascending air to escape, they prevent downward currents of cold air, *which would descend through the chimney*, and carry with them some part of the smoke and vapour of the fire.

600. *Why do chimneys smoke when fires are first lighted?*

Because the heat is for a time *insufficient to expand the air* and move the cold column of air within the chimney.

The first effect of a slight rarefaction of air at the base of the chimney is, to cause the column of cold air to press downwards, and therefore the smoke is forced into the room.

601. *Why does opening the window prevent the chimney from smoking?*

Because it supplies a free current of air, which, acting upon the warm air that is being produced by the fire, *assists to remove the cold volume* within the chimney.

602. *Why may the window be closed when the fire has increased?*

Because then the cold air within the chimney has been displaced, and there is a *sufficient upward draught* to carry the smoke away.

603. *Why will not opening the door answer as well as opening the window?*

Because the other fires burning within the house act upon the air which the house contains; and when a door is opened, air may be drawn *from a cold room* towards *a room in which there is a fire.* The door should therefore be closed, and the window opened.

604. *Why do most chimneys smoke in gusty weather?*

Because the sudden gusts of wind *disturb the relations existing* between the warm air of the chimney and the cold outer air. As the gusts sweep rapidly by the opening of the chimney top they remove suddenly the pressure of the cold air, and there is then a rush of air and smoke up the chimney. But when the gust ceases, the weight of the cold air being suddenly thrown upon the ascending column, checks its ascent, and drives it downwards.

605. *Why is there sometimes a smell of soot in rooms in summer time?*

Because the cool air of the chimney (having been confined therein for some time, and acquired the smell of the soot), being heavier than the warm air of the room, *descends* and diffuses a sooty smell.

CHAPTER XXIX.

606. *What is water?*

Water is a fluid composed of *two* volumes of *hydrogen* to *one* of *oxygen*, or *eight* parts by weight of *oxygen* to *one* of *hydrogen*. It is nearly colourless and transparent.

607. *Why does water become solid when it freezes?*

Because the *latent heat of the water* passes away from between its atoms into the air; the atoms, therefore, draw closer together.

608. *Why, if the atoms of water draw closer together when freezing, does ice expand, and occupy greater space than water?*

Because, *when the atoms of water are congealing*, they do not form a *compact mass*, but arrange themselves *in groups of crystal points*, which occupy greater space. Water *contracts* when freezing until it sinks to 40 deg., and then it *expands* as ice is formed.

32 deg. is said to be the *freezing* point, but it should be called the *frozen* point.

" As the hart panteth after the water-brooks, so panteth my soul after thee, O God."—PSALM XLII.

609. *Why does water boil?*

Because heat, *entering into the lower portions* of the water, *expands* it; the heated portions are then *specifically lighter* than those that are cooler; the hot water therefore *rises upward, and forces the cooler water down.*

610. *What proportion of the earth's surface is covered with water?*

There are about one hundred and forty-seven millions of square miles of *water* to forty-nine and a half millions of square miles of *land*.

611. *What is the amount of water pressure?*

The pressure of the sea, at the depth of 1,100 yards, is equal to 15,000 lbs. *to the square inch.*

612. *What element is the most abundant in nature?*

Oxygen, which forms so large a part of *water.* Of animal substances, *oxygen* forms *three-fourths;* of *vegetable substances* it forms *four-fifths;* of *mineral substances* it forms *one-half;* it forms *eight-ninths* of the *waters* and *one-fifth* of the *atmosphere;* and, aggregating the whole creation, from *one-half to two-thirds* consists of *oxygen.*

613 *In what ways does man use oxygen?*

Man *eats, drinks, breathes,* and *burns* it, in various proportions and combinations. It is estimated that *the human race* consume in those various ways, 1,000,000,000 lbs. daily; that the *lower animals* consume double that amount; and that, in the varied works of nature, no less than 8,000,000,000 lbs. of *oxygen* are used *daily.*

614. *Why does water dissolve various substances?*

Because the *atoms of water* are very minute; they therefore *permeate the pores,* or spaces, between the atoms of those bodies, and, *overcoming their attraction for each other,* cause them to separate.

"And God said, Let the waters under the heaven be gathered together unto one place, and let the dry land appear: and it was so."—GEN. I.

615. *Why does hot water dissolve substances more readily than cold?*

Because the *heat assists to repel the particles* of the substance undergoing solution, and *gives the water a freeer passage* between the atoms.

616. *Why is the sea salt?*

Because salt is a mineral which prevails largely in the earth, and which, *being very soluble in water*, is taken up by the ocean.

Lakes and rivers, also—even those that are considered fresh—hold in solution *some degree of saline matters*, which they contribute to the ocean.

As, in the evaporations from the sea, the salt remains in it, while the vapours fall as rain, and again wash the earth and carry some of its mineral properties to the ocean, *the greater saltness of the sea*, as compared with rivers, is accounted for.

By some persons the opinion is entertained that the sea has been *gradually getting salter* ever since the creation of the world. This, they say, arises from the evaporation of water free from salt, and the return of the water to the sea, taking with it salt from the land.

617. *What is the estimated amount of salt in the sea?*

The amount of common salt in the various oceans is estimated at 3,051,342 *cubic geographical miles*, or about five times more than the mass of the moutains of the Alps.

618. *What is the depth of the sea?*

The extreme depth has not, probably, been ascertained. But Sir James Ross took soundings about 900 miles west of St. Helena, whence he found the sea to be nearly *six miles in depth*. Now, if we take the height of the highest mountain to be five miles, the distance from that extreme rise of the earth, to the known depth of the sea, will be no less than *eleven miles*.

619. *Why does iron rust rapidly when wetted?*

Because the water contains so large a proportion of oxygen, some

"As in water face answereth to face, so the heart of man to man."—
PROVERBS XXVII.

of which combines with the iron, and forms *an oxide of iron*, which is *rust*.

620. *Why does stagnant water become putrid?*

Because the *large amount of oxygen* which it contains accelerates the decomposition of dead *animal and vegetable substances* that accumulate in it.

621. *Is there any danger in drinking water on account of the living animalcules which it contains?*

No danger arises from the *living creatures* in water; but *putrefactive* matters may produce serious diseases.

622. *What is the best method of guarding against impurities?*

By obtaining water from the purest sources, and by filtering it before drinking, by which nearly all extraneous matters would be *separated from it.*

CHAPTER XXX.

623. *What is the thermometer?*

The thermometer is an instrument in which *mercury* is employed to indicate *degrees of heat*. Its name is derived from two Greek words, meaning *heat measurer*.

624. *Why does mercury indicate degrees of heat?*

Because it *expands* rapidly with *heat*, and *contracts* with *cold*; and as it passes freely through small tubes, it is the most convenient medium for indicating *changes of temperature*.

625. *Why are there Reaumur's Thermometers and Fahrenheit's Thermometers?*

Because their inventors, after whom they are named, adopted a different system of *notation*, or *thermometrical marks*; and as their thermometers have been adopted by various countries and authors, it is now difficult to dispense with either of them.

"He sendeth out his word, and melteth them, he causeth his wind to blow, and the waters flow."—PSALM CXLVII.

THERMOMETER.

Reaumur. *Fahrenheit.*

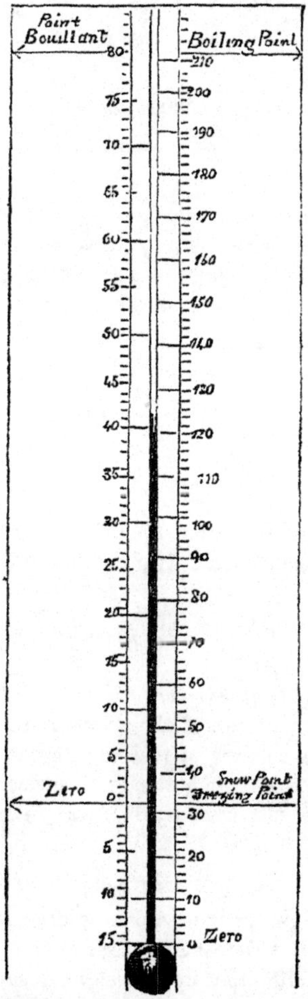

THE THERMOMETERS OF REAUMUR AND FAHRENHEIT COMPARED.

626. We have combined the two. The diagram will, we have no doubt, prove exceedingly useful to scientific readers and experimentalists. There is also another system of notation, adopted by the French, called the *centigrade*, but it is not much referred to in Great Britain. In the centigrade thermometer 0 zero is the freezing point, and 100 the boiling point. Fahrenheit's scale is generally preferred. Reaumur's is mostly used in Germany. Of Fahrenheit's scale 32 is the freezing point, 55 is moderate heat, 76 summer heat in Great Britain, 98 is blood heat, and 212 is the boiling point. Mr. Wedgwood has invented a thermometer for testing *high temperatures*, each degree of which answers to 130 degrees of Fahrenheit. According to his scale cast iron melts at 2,786 deg.; fine gold, at 2,016 deg.; fine silver, 1,873 deg.; brass melts at 1,869 deg.; red heat is visible by day at 980 deg.; lead melts 612 deg.; bismuth melts 476 deg.; tin melts 442 deg.; and there is a curious fact with regard to the three metals, lead, bismuth, and tin, that if they are mixed in the proportions of 5, 8, and 3 parts, respectively, the mixture (after previous fusion) will melt at a heat below that of boiling water.

627. *What is the difference between the thermometer and the barometer?*

In the thermometer the column of mercury is much smaller than in the barometer; and is sealed from the air; while in the barometer the column of mercury is open at one end to atmospheric influence.

628. *Why does the mercury in the thermometer, being sealed up, indicate the external temperature?*

"Behold, there ariseth a little cloud from the sea, of the bigness of a man's hand. And it came to pass, in the meantime, that the heaven was black with clouds and wind, and there was a great rain."—I. KINGS XVIII.

Because the heat passes through the glass, in which the mercury is enclosed, and *expanding or contracting the metal within the bulb*, causes the small column above it to *rise* or *fall*.

629. *When does the thermometer vary most in its indication of natural temperature?*

It varies more in the *winter* than in the *summer* season.

630. *Why does it vary more in the winter than in the summer?*

Because the temperature of our climate *differs more from the temperature of the torrid zones in the winter* than it does in the *summer*, and the *inequalities of temperature* cause frequent changes in the degree of prevailing heat.

The same remarks apply to the barometer.

CHAPTER XXXI.

631. *What is a barometer?*

A barometer is an instrument which *indicates the pressure of the atmosphere,* and which takes its name from two Greek words signifying *measurer of weight.*

632. *Why does a barometer indicate the pressure of the atmosphere?*

Because it consists of a tube containing *quicksilver*, closed at one end and open at the other, so that the pressure of the air upon the open end balances the weight of the column of mercury (quicksilver), and when the pressure of the air upon the open surface of the mercury increases or decreases, the mercury *rises or falls* in response thereto.

633. *Why is a barometer called also a "weather-glass?"*

Because changes in the weather are generally preceded by *alterations in the atmospheric pressure.* But we cannot perceive those changes as they gradually occur; the alteration in the height of the column of mercury, therefore, enables us to know that atmospheric changes are taking place, and, by observation, we are enabled to determine certain rules by which *the state of the weather may* be *foretold* with considerable probability.

634. *Why are barometers constructed with circular dials, and an index to denote changes?*

Because that is a convenient mechanical arrangement, by which the alterations of the relative pressures of the air and the mercury are *more clearly denoted than by an inspection of the mercury itself.*

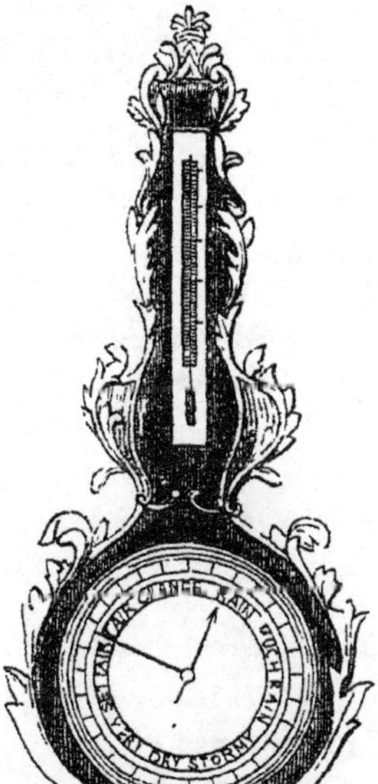

Fig. A.—BAROMETER.

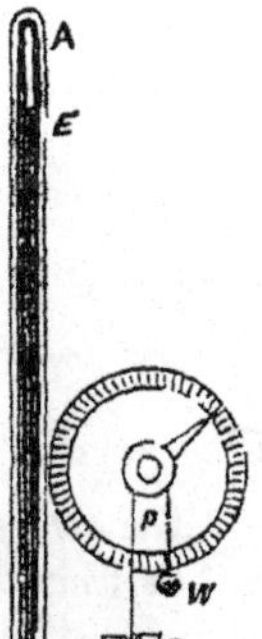

Fig. B.—TUBE OF BAROMETER, WHEEL, AND PULLEY.

Fig. B illustrates the mechanism at the back of the barometer. A is a glass tube; between A and E there exists a *vacuum*, caused by the weight of the mercury pressing downwards. This space being a vacuum makes the

"Fire and hail; snow and vapour; stormy wind fulfilling his word."—
PSALM CXLVIII.

barometrical column more sensitive, as there is no internal force to resist or modify the effects of the external pressure. E represents the height of the column of mercury, C the open end of the tube; F the weight resting on the surface of the mercury, P the pivot over which the string passes, and upon which the hand turns: W the weight which forms the pulley with the weight F.

635. *Why does the hand of the weather dial change its position when the column of mercury rises or falls?*

Because a weight, which *floats upon the open surface of the mercury*, is attached to a string, having a nearly equal weight at the other extremity; the string is laid over a revolving pivot to which the hand is fixed, and *the friction of the string turns the hand as the mercury rises or falls.*

636. *Why does tapping the face of the barometer sometimes cause the hand to move?*

Because the weight on the surface of the mercury frequently *leans against the sides of the tube,* and does not move freely. And, also, the mercury clings to the sides of the tube by *capillary attraction;* therefore, tapping on the face of the barometer *sets the weight free.* and overcomes the attraction which *impedes the rise or fall of the mercury.*

637. *Which is the heavier, dry or vaporised air?*

Dry air is *heavier* than air impregnated with vapours.

638. *Why is dry air heavier than moist air?*

Because of the *extreme tenuity of watery vapours,* the density of which is *less than that of atmospheric air.*

639. *Why does the fall of the barometer denote the approach of rain?*

Because it shows that as the air *cannot support the full weight of the column of mercury,* the atmosphere must be thin with watery vapours.

The fall of the mercury in the long arm of the tube would cause the weight F to be pressed upwards. This would release the string to which the weight W is attached; it would, therefore fall, and turn the hand down to Rain, or Much Rain.

"He gathereth the waters of the sea together as an heap; he layeth up the depth in storehouses."—PSALM XXXIII.

640. *Why does the rise of the barometer denote the approach of fine weather?*

Because the external air becoming dense, and free from highly elastic vapours, pressed with increased force upon the mercury upon which the weight F floats; that weight, therefore, sinks in the short tube as the mercury rises in the long one, and, in sinking, turns the hand to Change, Fair, &c.

641. *Why does the barometer enable us to calculate the height of mountains?*

Because, as the barometer is carried up a mountain, *there is a less depth of atmosphere above to press upon the mercury;* it therefore falls, and by comparing various observations, it has been found practicable to *calculate the height of mountains by the fall of the mercury in a barometer.*

642. *To what extent of variation is the weight of the atmosphere liable?*

It may vary as much as *a pound and a half to the square inch* at the level of the sea.

643. *When does the barometer stand highest?*

When there is a *duration of frost*, or when *north-easterly winds* prevail.

644. *Why does the barometer stand highest at these times?*

Because the atmosphere is exceedingly *dry and dense*, and fully balances the *weight of the column of mercury.*

645. *When does the barometer stand lowest?*

When *a thaw follows a long frost;* or when *south-west winds* prevail.

646. *Why does the barometer stand lowest at these times?*

Because *much moisture exists in the air*, by which it is rendered less dense and heavy.

"But the land, whither ye go to possess it, is a land of hills and valleys, and drinketh water of the rain of heaven."—DEUT. XI.

647. *What effect has heat upon the barometer?*

It causes the mercury to fall, *by evaporating moisture into the air.*

648. *What effect has cold upon the barometer?*

It causes the mercury to rise, by *checking evaporation*, and *increasing the density of the air.*

649. In noting barometrical indications, more attention should be paid to the *tendency* of the mercury at the time of the observation than to the *actual state of the column*, whether it stands *high* or *low*. The following rules of barometic reading are given as generally accurate, but liable to exceptions:—

Fair weather, indicated by the *rise* of the mercury.

Foul weather, by the *fall* of the mercury.

Thunder, indicated by the *fall* of the mercury in *sultry weather*.

Cold, indicated by the *rise* of the mercury in spring, autumn, and winter.

Heat, by the *fall* of the mercury in summer and autumn.

Frost, indicated by the *rise* of the mercury in winter.

Thaw, by the *fall* of the mercury during a frost.

Continued bad weather, when the *fall* of the mercury has been *gradual* through several fine days.

Continued fine weather, when the *rise* of the mercury has been *gradual* through several foul days.

Bad weather of short duration, when it sets in quickly.

Fine weather of short duration, when it sets in quickly.

Changeable weather, when an *extreme* change has *suddenly* set in

Wind, indicated by a rapid *rise* or *fall unattended by a change of temperature.*

The mercury *rising*, and the air becoming *cooler*, promises *fine weather*; but the mercury *rising*, and the air becoming *warmer*, the weather will *be changeable.*

If the top of the column of mercury appear *convex*, or curved upwards, it is an additional proof that the mercury is *rising*. Expect *fine* weather.

If the top of the column is *concave*, or curved downwards, it is an additional proof that the mercury is *falling*. Expect *bad* weather.

CHAPTER XXXII.

650. *What is sound?*

Sound is an *impression produced upon the ear* by *vibrations* of *the air.*

651. *What causes the air to vibrate and produce sounds?*

The atoms of *elastic bodies* being caused to *vibrate* by the application of some kind of force, *the vibrations of those atoms are imparted to the air*, and sound is produced.

652. *How do we know that sounds are produced by the vibrations of the air, induced by the vibrations of the atoms of bodies?*

If we take a tuning fork, and hold it to the ear, we hear *no sound.* If we move it rapidly through the air, or if we blow upon it, it produces *no sound;* but if we *strike it, a sound immediately occurs;* the vibration of the fork may be seen, and felt by the hand that holds it; and, *as those vibrations cease*, the sound *dies away.*

653. *How do we know that without air there would be no sound?*

Because if a tuning-fork were to be struck in a *vacuum* (as under the receiver of an air pump) *no sound* would be heard, although the *vibrations* of the fork could be *distinctly seen.*

654. *How are the vibrations of sonorous bodies imparted to the air?*

When a bell is struck, the force of the blow gives an instant agitation to all its particles. The air around the bell is driven back by the impulse of the force, and thus a *vibration of compression* is imparted to the air; but the air returns to the bell, by its own natural elasticity, thus producing a *vibration of expansion* when it is again struck, and thus *successive vibrations* of compression and expansion are *transmitted through the air.*

655. *How rapidly are these vibrations transmitted through the air?*

They travel at a rate of rather more than *a quarter of a mile in a second*, or *twelve miles and three-fourths in a minute.*

656. *Do all sounds travel at the same rate?*

All sounds, whether strong or weak, high or low, musical or discordant, *travel with the same velocity.*

"And I will cause the noise of thy songs to cease, and the sound of thy harps shall no more be heard."—EZEKIEL XXVI.

657. *Why are bells and glasses stopped from ringing by touching them with the finger?*

Because the contact of the finger *stops the vibration* of the atoms of the metal and glass, which therefore *cease to impart vibrations to the air.*

658. *Why does a cracked bell give discordant sounds?*

Because the *connection* between the atoms of the bell being *broken,* their vibrations are not uniform: some of the atoms vibrate *more intensely* than the others; the vibrations imparted to the air are, therefore, *jarring and discordant.*

659. *Why, when we see a gun fired at the distance, do we see the flash and smoke before we hear the report?*

Because *light,* which enables us to *see,* travels at the velocity of 192,000 miles in a *second;* while *sound,* by which we *hear,* travels only at the rate of a quarter of a mile in a *second.*

660. *Why does the tread of soldiers, when marching in long ranks, appear to be irregular?*

Because the sounds proceeding from *different distances* reach our ears in *varying periods of time.*

661. *Why does the length of a wire or string determine the sound that it produces?*

Because the *shorter the string* the *more rapid* are its vibrations when struck.

662. *Why does the tension of a wire or string affect its vibrations?*

Because when the string or wire is tight, a touch communicates vibrations to *all its particles;* but when it is loose the vibrations are *imperfectly communicated.*

663. *Why are some notes low and solemn, and others high and quick?*

Because the vibrations of musical strings vary from 32 vibrations in a second, which produces a soft and deep bass, to 15,000 vibrations in a second, which produces the sharpest treble note.

"My heart maketh a noise in me: I cannot hold my peace, because thou hast heard, O my soul, the sound of the trumpet, the alarm of war."—JEREMIAH. IV.

664. *What are the numbers of vibrations in a second that produce the various musical sounds?*

C or Do, 480 vibrations in a second; B or Si, 450 vibrations; A or La, 400 vibrations; G or Sol, 360 vibrations; F or Fa, 320 vibrations; E or Mi, 300 vibrations; D or Re, 270 vibrations; It is thus seen that the *more rapid* the vibrations, the *higher* the note, and *vice versa*.

665. *Why can our voices be heard at a greater distance when we speak through a tube?*

Because the vibrations are *confined to the air within the tube*, and are not interfered with by *other vibrations* or movements in the air; the tube itself is also a *good conductor of sound*.

666. *Is air a good conductor of sound?*

Air is a *good conductor*, but water is a *better conductor than air;* wood, metals, the earth, &c., are also *good conductors*.

667. *Why can we hear sounds at a greater distance on water than on land?*

For various reasons: because the smooth surface of water is a good conductor; because there are fewer noises, or counter vibrations, to interfere with the transmission of sound; and because there are no elevated objects to impede the progress of the vibrations.

668. *Why do sea-shells give a murmuring noise when held to the ear?*

Because what may be called *expended vibrations* always exist in air where various sounds are occuring. These *tremblings* of the air are received upon the thin covering of the shell, and thus being *collected into a focus*, are *transmitted to the ear*.

669. *Why can people in the Arctic regions converse when more than a mile apart?*

Because there the air, being *cold and dense*, is a very good conductor; and the *smooth surface of the ice* also favours the *transmission of sound*.

"The morning is come unto thee, O thou that dwellest in the land; the time is come, the day of trouble is near, and not the sounding again of the mountains."—EZEKIEL VII.

670. *Why do savages lay their heads upon the earth to hear the sounds of wild beasts, &c.?*

Because the earth is a good conductor of sound. For this reason, also, persons *working under ground in mines* can hear each other digging at considerable distances.

671. *Why can church clocks be heard striking much more clearly at some times than at others?*

Because the density of dry air improves the *sound-conducting power* of the atmosphere. The transmission of sounds is also assisted by the direction of the winds.

672. *Why may the scratching of a pin at one extremity of a long pole be heard by applying the ear to the opposite extremity?*

Because wood is a good conductor of sound, and its atoms are *susceptible of considerable vibration.* It is, therefore, chosen in numerous instances for the construction of *musical instruments.*

Deaf persons have been known to derive pleasure from music by placing their hands upon the wood work of musical instruments while being played upon.

673. *Why is the hearing of deaf persons assisted by ear-trumpets?*

Because ear-trumpets *collect the vibrations of the air* into a focus, and make the sounds produced thereby more intense.

674. *Why are sounding-boards used to improve the hearing of congregations?*

Because, being suspended over, and a little behind, the speaker, they *collect the vibrations* of the air, and *reflect* them towards the congregation.

675. *What are echoes?*

Echoes are sounds *reflected* by objects on which they strike.

676. *Why do some echoes occur immediately after a sound?*

Because the reflecting surface is *very near;* therefore the sound returns immediately.

"And even things without life giving sound, whether pipe or harp, except they give a distinction in the sounds, how shall it be known what is piped or harped?"—CORINTH. XIV.

677. *Why do some echoes occur a considerable time after a sound?*

Because they are at a considerable distance, and the sound takes time to travel to it and time to return.

678. *Why do some echoes change the tone and quality of sound?*

Because the reflecting surface, having vibratory qualities of its own, *mingles its own vibrations with those of the sound.*

679. *Why are there sometimes several echoes to one sound?*

Because there are various *reflecting surfaces* at different distances, each of which returns an echo.

680. *Are sounds reflected only by distant objects?*

Sounds are doubtless reflected by *walls and ceilings* around us. But we do not perceive the echoes, because they are so near that they occur at the same moment with the sound. In lofty buildings, however, there is frequently a *double sound*, making the utterance of a speaker indistinct. This arises from the echo following very closely upon the sound.

681. *Why, when we are walking under an arch-way or a tunnel, do our voices appear louder?*

Because the sounds of our voices are *immediately reflected.* And as a *gas reflector increases the intensity of light*, so *a sound reflector* will *increase the apparent strength of our voices.*

682. There are many places where remarkable echoes occur. On the banks of the Rhine, at Lurley, if the weather be favourable, the report of a rifle, or the sound of a trumpet, will be repeated at different periods, and with various degrees of strength, from crag to crag, on opposite sides of the river alternately. A similar effect is heard in the neighbourhood of some of the Lochs in Scotland. There is a place at Woodstock, in Gloucestershire, which is said to echo a sound fifty times. Near Roseneath, a few miles from Glasgow, there is a spot where, if a person plays a bar of music upon a bridge, the notes will be repeated by an echo, but a third lower; after a short pause, another echo is heard, again in a lower tone; then follows another pause, and a third repetition follows in a still lower key. The effect is very enchanting. The whispering galleries of St. Paul's, of the cathedral church of Gloucester, and of the Observatory of Paris, owe their curious effects to those laws of the reflection of sound by which echoes are produced; but in these cases the effect is assisted by the eliptical form of the edifice, each person being in the focus of an elipse.

"Where no wood is, there the fire goeth out: so where there is no tale-bearer the strife ceaseth."—PROVERBS XXVI.

CHAPTER XXXIII.

683. *What are gases?*

Gases are permanently elastic aeriform fluids, or substances which have the appearance of air; they are transparent, elastic, ponderable, and, with few exceptions, are invisible.

684. *Why may gases be obtained from solid bodies?*

Because caloric enters into combination with the substances forming the gas, and produces therein a high degree of elasticity.

685. *Why is not vapour considered to be gas?*

Because, although it resembles gas in many respects, the caloric which produces it is but feebly combined with its particles, and it exhibits a tendency to part with the caloric, and return to its previous state of water. But in gases caloric is united by a very forcible affinity to the substance forming the gas, and produces permanently aeriform elastic fluids.

686. *Why are gases called variously hydrogen, oxygen, carbonic acid, &c.?*

Because the gases are named after the solids that form their base: thus, oxygen gas is oxygen rendered gaseous by caloric; and the same explanation applies to hydrogen, nitrogen, carbonic acid, and other gases.

687. *Why is carbonic acid gas so called?*

Because it is formed of a combination between carbon and oxygen.

688. Acids are distinguished according to the proportion of oxygen which they contain. The termination *ic* and *ous*, are employed to denote the larger or smaller proportion of oxygen, as nitr*ic* acid, sulphur*ic* acid, nitr*ous* acid, and sulphur*ous* acid. The termination *ic* expresses the larger proportion of oxygen.

689. *What are the properties of oxygen gas?*

The oxygen gas in the air is the chief supporter of combustion, and the vehicle of heat. It is absolutely necessary to animal life. Oxygen performs an important part in most of the changes which take place in the mineral, animal, and vegetable kingdoms.

"Thus saith the Lord, Let not the wise man glory in his wisdom, neither let the mighty man glory in his might, let not the rich man glory in his riches."—JEREMIAH IX.

690. *What is oxygen?*

Oxygen is one of the most widely diffused of the elementary substances. It is a gaseous body.

691. *Why do persons who are walking or riding feel warmer than when they are sitting still?*

Because as they breathe more rapidly, the combustion of the *carbon* in the blood is increased by the *oxygen* inhaled, and greater heat is developed.

692. *Why does the fire burn more brightly when blown by a bellows?*

Because it receives with every current of air, a fresh supply of *oxygen*, which unites with the *carbon* and *hydrogen* of the coals, causing more rapid combustion and increased heat.

693. *Why does not the oxygen of the air sometimes take fire?*

Because oxygen, *by itself*, is incombustible. The wick of a candle, which retains the slightest spark, being immersed in oxygen, will instantly burst into a brilliant flame; and even a piece of iron wire made red-hot, and dipped in oxygen, will burn rapidly and brilliantly. Oxygen, though non-combustible of itself, is the most powerful *supporter of combustion*.

694. *Why do we know that oxygen will not burn of itself?*

Because when we immerse a burning substance into a jar of oxygen, it immediately burns with intense brilliancy; but directly it is withdrawn from the oxygen, the intensity of the flame diminishes, and the oxygen which remains is *unaffected*.

695. *Why do we know that oxyyen is necessary to our existence?*

Because animals placed in any kind of gas, or in any combination of gases, where oxygen *does not exist*, die in a very short time.

"Surely every man walketh in a vain show, surely they are disquieted in vain: he heapeth up riches, and knoweth not who shall gather them."—PSALM XXXIX

696. *Where is oxygen found?*

It is found in the air, mixed with *nitrogen;* in water combined with *hydrogen;* in the tissues of vegetables and animals; in our blood; and in various compounds called, from the presence of oxygen, *oxides.*

697. *Why is the oxygen of the air mixed so largely with nitrogen?*

Because *oxygen* in any greater proportion than that in which it is found in the atmosphere would be too exciting to the animal system. Animals placed in *pure oxygen* die in great agony from fever and excitement, amounting to madness.

698. *What is nitrogen?*

Nitrogen is an elementary body in the form of gas; it is the principal of *nitre* made gaseous by combination with caloric.

699. *Where is nitrogen found?*

It is chiefly found in the air, of which it constitutes 79 out of 100 volumes. It may be mixed with oxygen in various proportions; but in the atmosphere it is uniformly diffused. It is found in most animal matter, *except fat and bone.* It is not a constituent of the *vegetable acids,* but it is found in most of the *vegetable alkalies.*

700. *What are acids?*

Acids are a numerous class of chemical bodies. They are generally sour. Usually (though there are exceptions) they have a great affinity for water, and are easily soluble therein; they unite readily with most *alkalies,* and with the various *oxides.* All acids are compounds of two or more substances. Acids are found in all the kingdoms of nature.

701. *What are alkalies?*

Alkalies are a numerous class of substances that have a great affinity for, and readily combine with, *acids,* forming *salts.* They exercise peculiar influence upon vegetable colours, turning blues green, and yellows reddish brown. But they will restore the colours of vegetable blues which have been reddened by acids; and, on the other hand, the *acids* restore vegetable colours that have been altered by the *alkalies*. Alkalies are found in all the kingdoms of nature.

702. *Could animals live in nitrogen*

No; they would immediately die. But a mixture of *oxygen* and *nitrogen*, in equal volumes, constitutes *nitrous oxides*, which gives a pleasurable excitement to those who inhale it, causing them to be merry, almost to insanity; it has, therefore, been called *laughing gas*.

703. *Why does nitrous oxide produce this effect?*

Because it introduces into the body more *oxygen* than can be consumed. It, therefore, deranges the nervous system, and being a powerful stimulant, gives an unnatural activity to the nervous centres and the brain.

704. *In what proportions are the atmospheric gases found in the blood?*

The mean quantity of the gases contained in the human blood has been found to be equal to 1-10th of its whole volume. In *venous* blood, the average quantity of *carbonic acid* is about 1-18th, that of *oxygen* about 1-85th, and that of *nitrogen* about 1-100th of the volume of the blood. In *arterial* blood their quantities have been found to be *carbonic acid* about 1-14th, *oxygen* about 1-38th, and *nitrogen* about 1-72nd.

705. *When is nitrogen taken into the blood from the air?*

Such a supposition is highly improbable. It is probably derived from *nitrogenised food*, just as *carbonic acid* is derived from *carbonised food*.

706. *What is venous blood?*

Venous blood is that which returns through the *veins* of the body from the organs to which it has been circulated.

707. *What is arterial blood?*

Arterial blood is that which flows from the heart through

"And the strong shall be as tow, and the maker of it as a spark, and they shall both burn together, and none shall quench them."—ISAIAH I.

the *arteries* to nourish the parts where those arteries are distributed.

708. *What is the difference between venous and arterial blood?*

Venous blood contains *more* carbonic acid, and *less* oxygen and nitrogen than arterial blood.

CHAPTER XXXIV.

709. *Will nitrogen burn?*

It will not burn, nor will it support combustion.

710. *What is the difference between "burning" and "supporting" combustion?*

Oxygen gas will not burn of itself, but it aids the decomposition by fire of bodies that are combustible. It is therefore called a *supporter of combustion.* But hydrogen gas, *though it burns of itself,* will extinguish a flame immersed in it. It is therefore said to be a body which will *burn,* but *will not support combustion.*

711. *What becomes of the nitrogen that is inhaled with the air?*

It is thrown off with the breath, mixed with *carbonic acid gas,* and flies away to be renewed by a fresh supply of oxygen.

712. *Where does nitrogen find a fresh supply of oxygen?*

In the atmosphere. Nitrogen is said to possess a remarkable tendency to *mix* with oxygen, without having a positive chemical *affinity* for it. That is to say, neither the *oxygen* nor the *nitrogen* undergo any change by the union, except that of *admixture.* The oxygen and the nitrogen still possess their own peculiar properties. Oxygen and nitrogen are found in nearly the same proportions in all climates, and at all altitudes.

713. *Why is nitrogen also called azote?*

It is so called from the Greek, on account of its destructive effects upon animal life when breathed in a pure state.

"When his candle shined upon my head, and when by his light I walked through darkness."—JOB XXIX.

714. *In combustion does any other result take place besides the union of oxygen and carbon forming carbonic acid gas?*

Yes. Usually *hydrogen* is present, which, in burning, unites with *oxygen,* and forms *water.*

715. *What is carbon?*

Carbon is one of the elementary bodies. It abounds throughout the vegetable kingdom, and is also contained in animal and mineral bodies. It is commonly known in the forms of charcoal, coal, soot, &c., and the most valuable of all gems, the diamond, is nothing but pure crystalised carbon.

716. Hence we derive another of the beautiful lessons of science—a lesson which teaches us to *despise nothing that God has given.* The soot which blackens the face of a chimney-sweep, and the diamond that glistens in the crown of the monarch, consist of the same element in merely a different atomic condition. What a lesson of humility this teaches to Pride! The haughty beauty as she walks the ball-room, inwardly proud of the radiance of her gems as they rise and fall upon her breast, little thinks or knows that *every breath that is expired around her wafts away the like element of which her treasures are composed.* That even in our flesh and bones the same abounding substance lies hid; and that the buried tree of the primitive world, and the little flower of to-day, are both instrumental in giving this singular element to man!

717. *Is charcoal a simple or a compound substance?*

It is a compound substance, consisting of the woody fibre of vegetables partially oxidised; it also contains a small proportion of hydrogen, and some earthy matter.

718. *What is the use of carbon in the vegetable kingdom?*

Carbon forms the chief part of the solid bases of *all vegetables,* from the most delicate flower in the garden to the huge oak of the forest.

719. *What is the use of carbon in the animal kingdom?*

As it enters into all vegetable, so it forms a part of *all animal bodies.*

720. *What are the proportions of oxygen and carbon that form carbonic acid gas?*

One hundred parts of carbonic acid gas consist of *twenty-eight* parts of *carbon* and *seventy-two* of *oxygen.*

"Lord make me know mine end, and the measure of my days, that I may know how frail I am."—PSALM XXXIX.

721. *Why does this mixture, called carbonic acid gas, prove fatal to life?*

Because the carbon has such *a powerful affinity for oxygen* that it will not impart it to the blood. Animals that breathe carbonic acid gas become *suffocated* for want of *oxygen.*

722. *In what state does carbonic acid exist in nature?*

In gas, in mixture with air and water, and in various chemical combinations.

723. *Why will a lighted candle placed under an inverted glass cease to burn after a few minutes?*

Because in burning the oxygen of the air combines with the carbon of the candle, and forms *carbonic acid gas,* which will neither support flame nor animal life.

724. *Why does carbonic acid gas accumulate on the floors of places where it is in excess?*

Because it is the *heaviest of all gases;* it therefore has a tendency to fall and become concentrated in low places.

725. *Why is it proper to lower a lighted candle into wells and deep places before venturing into them?*

Because frequently there is in such places an accumulation of *carbonic acid gas,* which would be instantly fatal to life. If the candle should cease to burn, it would show that a person could not descend until the pit or well had been ventilated, without loss of life.

726. *Why would the going out of the candle show the danger that might attend a person entering the well?*

Because oxygen is the gas which sustains life as well as flame; and where a flame will not burn, animals cannot live.

727. *Why do dogs that enter the Grotto del Cano in Italy, drop down dead?*

Because a *stream of carbonic acid* gas flows along the floor of the grotto, from the sources of which it is generated in the earth.

728. *Why may men enter the cave without feeling the effects of the poisonous gas?*

Because the gas, *being heavy*, lies upon the floor of the cave; the head of a man rises above it, and he breathes the air.

729. *Why should quick-lime be thrown into a well charged with carbonic acid?*

Because the lime, if sprinkled with a little water, would begin to slacken, and as it slackened it would *absorb the carbon of the gas*, forming carbonate of lime. Persons might afterwards descend in safety.

730. *Why do fatal accidents often happen from burning charcoal in chambers?*

Because, wherever charcoal is burnt, *carbonic acid gas is formed abundantly*, by the combination of the oxygen of the air with the carbon of the charcoal.

731. *Why do workmen sometimes lose their lives by sleeping too near lime kilns?*

Because limestone is a carbonate of lime, and when it is being burned large volumes of *carbonic acid gas* are driven off. Persons who lie down near kilns are likely to be involved in streams of this gas, and to be suffocated.

CHAPTER XXXV.

732. *Why does water contain carbonic acid gas?*

Because carbonic acid gas is *absorbable by water*. If a jar be partly filled with the gas, and it be allowed to stand some hours over water, an absorption will take place till none of the gas remains.

733. *Why has fresh water a brisk and pleasant taste?*

Because it contains *carbonic acid gas*.

734. *Why does water loose its brisk and pleasant taste when boiled?*

Because it is wholly deprived of *carbonic acid gas* by boiling.

735. *Why does soda-water effervesce and sparkle?*

Because it is artificially impregnated with *carbonic acid gas*. By pressure, water may be combined with twice and a half of its

"As coals are to burning coals, and wood to fire; so is a contentious man to kindle strife."—PROVERBS XXVI.

own bulk of this gas. But as soon as the pressure is removed, *the gas escapes.*

736. *Why do cyder, perry, ale, champagne, &c., sparkle and effervesce?*

Because they contain *carbonic acid gas*, a great part of which, on uncorking the bottle, expands again to the gaseous form. In escaping, it produces the froth which runs so abundantly from bottles.

737. *Why is carbonic acid gas produced during fermentation?*

Because in vinous fermentation a decomposition of the sweet matter takes place, in which a part of the oxygen is disengaged; this unites with a part of the carbon of the sugar, forming *carbonic acid gas.*

738. *What is hydrogen?*

Hydrogen is the base of what has been generally called inflammable air, now known as *hydrogen gas;* it is also one of the component parts of water; but it does not exist except in combination with other substances, or in the *gaseous form*, alternated by *caloric.*

739. *Why do balloons filled with hydrogen gas ascend?*

Because hydrogen gas is lighter than common air. Pure hydrogen gas is *the lightest* of all known bodies.

740. *Why is hydrogen, the lightest of all gases, spoken of also as a solid body?*

Because, although when in the aerial state, it is the lightest of all known substances, yet when imbibed by living vegetables it becomes *a solid*, and forms oil, wax, resin, &c.; and in combination with oxygen it constitutes water, which has the property of becoming either solid, fluid, or aeriform.

741. *Will hydrogen support combustion?*

Although it will burn, yielding a feeble green light, it will, if pure, extinguish a flame that may be immersed in it. Hydrogen will therefore *burn*, but will not *support combustion.*

742. *Why will hydrogen explode, if it will not support combustion?*

"But the wise answered saying, Not so; lest there be not enough for us and you: but go ye rather to them that sell, and buy for yourselves."—MATT. XXV.

When hydrogen explodes it is always in combination with *oxygen*, or with the common air, which contains *oxygen*. *Two* measures of hydrogen and *one* of oxygen form a most explosive compound.

743. *Why does hydrogen explode, when mixed with oxygen, upon being brought in contact with fire?*

Because of its strong affinity for *oxygen*, with which, upon the application of heat, it unites to form water.

744. *What is combustion?*

Combustion is a chemical process by which certain substances *decompose oxygen gas*, absorb its base, and suffer its caloric to escape, producing heat.

When a combustible body is heated to a high degree, it possesses such an attraction for oxygen, that it *absorbs it for the air*, and fixes it in a *solid* form; while the caloric which gave to the oxygen its aeriform condition is *set free and diffused.*

745. *Whence comes the light that is produced by combustion?*

It is supposed that light, like heat, is latent in those bodies that evolve it during chemical action. If this be true, oxygen gas supplies the *heat* and some of the *light* that is given out in burning.

746. *What effects are generally produced by the process of combustion?*

The effects of combustion are in some degree modified by the nature of the body burned.

In the burning of a common fire, *carbonic acid gas* is produced, by the combination of the *carbon* of the coals and the *oxygen* of the air; and *water* is formed by the combination of the *hydrogen* of the coals with other portions of *oxygen.*

Hence ordinary combustion consists of a compound chemical action, during which *carbonic acid gas* and *water* are formed, and *light* and heat liberated.

747. *Why is there such a continuous heat given out of a fire?*

Because a *fresh supply* of oxygen is conveyed to the fire in the currents of air that move towards it.

It is from the oxygen that the heat chiefly emanates; hence the heat *will continue* as long as the combustible body lasts and maintains a temperature sufficiently high to *decompose oxygen gas.*

748. *Where does hydrogen chiefly exist?*

In the form of *water,* where it exists in combination with *oxygen. Eleven* parts of hydrogen, and *eighty-nine* of oxygen, form water.

749. *Is hydrogen found elsewhere?*

It is never found but in a state of combination; united with oxygen, it exists in *water;* with nitrogen, in *ammonia;* with chlorine, in *hydro-chloric acid;* with fluorine, in *hydrofluoric acid;* and in numerous other combinations.

750. *Is the gas used to illuminate our streets, hydrogen gas?*

It is; but it is combined with carbon, derived from the coals from which it is made. It is therefore called *carburetted hydrogen,* which means *hydrogen* with *carbon.*

751. *How is hydrogen gas obtained from coals?*

It is driven out of the coals by heat, in closed vessels, which prevents its union with *oxygen.*

752. *What becomes of the water which is formed by the burning of hydrogen in oxygen?*

It passes into the air in the form of watery vapour. Frequently it condenses, and may be seen upon the walls and windows of rooms where many lights or fires are burning. Sometimes, also, portions of it become condensed in the globes of the glasses that are suspended over the jets of gas. *A large volume of these gases forms only a very small volume of water.*

753. *What becomes of the carbonic acid gas which is produced by combustion?*

It is diffused in the air, which should be removed by adequate ventilation.

754. *What proportion of carbonic acid gas is dangerous to life?*

Any proportion over the natural one of 1 per cent. may be

regarded as *injurious*. But it has been ascertained that *five per cent.* of carbonic acid gas in the atmosphere is *dangerous* to life.

755. *Is an escape of hydrogen gas from a gas-pipe dangerous to life?*

It is dangerous, first, by *inhalation*. There are no less than six deaths upon record of persons who were killed by sleeping in rooms near to which there was a leakage of gas.

It is dangerous, secondly, by *explosion*.

756. In 1848, an explosion of gas occurred in Albany-street, Regent's-park, London. The gas accumulated in a shop for a very short time only. It had been escaping from a crack in the meter for about one hour and twenty minutes. The area of the room was about 1,600 cubic feet. When the gas exploded, it blew out the entire front of the premises, carried two persons through a window into an adjoining yard, and forced another person on to the pavement on the opposite side of the street, where she was killed. The effect of the explosion was felt for more than a quarter of a mile on each side of the house, and most of the windows in the neighbourhood were shattered. The iron railings over the area of the house directly opposite were snapped asunder; and a part of the roof, and the back windows of another house, were carried to a distance of from 200 to 300 yards. The pavement was torn up for a considerable length, and the damage done to 103 houses was afterwards reported to amount to £20,000. Other serious explosions have taken place. The explosions of "*coal damp*," which frequently occur in mines, are of a similar character.

757. *What proportion of hydrogen gas with atmospheric air will explode?*

According to the researches of Sir Humphrey Davy, *seven* or *eight* parts of *air*, to *one* of *gas*, produce the greatest explosive effect; while *larger* proportions of gas are less dangerous. A mixture of *equal parts* of gas and air will burn, but it will not explode, The same is the case with a mixture of *two* of *air*, or *three* of *air*, and *one* of *gas*; but *four* of *air* and *one* of *gas* begin to be explosive, and the explosive tendency increases up to *seven* or *eight* of *air* and *one* of *gas*, after which the increased proportion of gas diminishes the force of the explosion.

758. *What is the best method of preventing the explosion of gas?*

Observe the rule, *never to approach a supposed leakage with a light.* Fortunately the gas, which threatens our lives, warns us of the danger by its pungent smell. The first thing to be done is to open windows and doors, and to ventilate the apartment. Then turn the gas off at the main, and wait a short time until the accumulated gas has been dispersed.

759. *Does hydrogen gas rise or fall when it escapes?*

Being *twelve times lighter than common air* it *rises*, and therefore it would be better for ventilation to open the window at the *top* than at the *bottom*. But all gases exhibit a strong tendency to *diffuse themselves*, and therefore they do not rise or fall in the degree that might be anticipated.

760. *What proportion of hydrogen in the air is dangerous to life, if inhaled?*

One-fiftieth part has been found to have a *serious effect* upon animals. The effects it produces upon the human system are those of depression, headache, sickness, and general prostration of the vital powers. It is therefore advisable to observe precautions in the use of gas.

761. *What proportion of gas in the air may be recognised by the smell?*

By persons of acute powers of smelling it may be recognised when there is *one* part of *gas* in *five hundred parts of atmospheric air*; but it becomes very perceptible when it forms *one* part in *a hundred and fifty*. Warning is, therefore, given to us long before the point of danger arrives.

762. *What other sources of hydrogen are there in our dwellings?*

It arises from the decomposition ot animal and vegetetable substances, containing *sulphur* and *hydrogen*. These give off a gas called *sulphuretted hydrogen*, from which the fætid effluvium of drains and water-closets chiefly arises. We should, therefore, take every precaution to secure effective drainage, and to keep drain-traps in proper order.

763. *May the use of gas for purposes of illumination be considered highly dangerous?*

Not if it is intelligently managed. The appliances for the regulation of gas are so very simple and perfect, that accidents seldom arise except from neglect. In England 6,000,000 tons of coal are usually consumed in the manufacture of gas, producing 60,000,000,000 cubic feet of gas. And yet accidents are of very uncommon occurrence.

764. *What becomes of the oil that is burnt in lamps?*

The carbon of the oil unites with the oxygen of the atmosphere, and forms *carbonic acid gas;* while its *hydrogen* unites with another portion of oxygen, and forms *water*. Every 100 ounces of oil thus burnt produce 130 ounces of water.

765. *Are any substances annihilated by being burnt?*

When bodies are *burnt* they are not *destroyed;* they had previously formed together one kind of compound, and they now separate from each other, at the high temperature to which they are exposed, in order to form others with the vital air in contact with them; and such of the principles as cannot unite with the vital air—viz., the earth, and some saline or metallic particles, compose the *cinder*.

766. The economy of nature should teach us a very impressive lesson—*nothing is suffered to be wasted*, not even the slightest atom. As soon as any body has fulfilled its purpose in one state of being, it is passed on to another. The candle, existing no longer as a candle, is flying upon the wings of the air as *carbonic acid gas*, and as *water*. These probably find their way to the garden or the field, where the carbonic acid gas forms *the food of the plant*, and the water affords a refreshing *drink*. And can it be supposed that the Almighty Being, who has thus economised the existence of the *material* creation, should be less mindful of the immaterial *soul* of man? There is an eternity before us, the certainty of which is evidenced by even by the laws of the material creation.

CHAPTER XXXVI.

767. *Why is it necessary to apply a flame to light a candle?*

Because the heat of *flame* is greater than that of a *red heat;* when flame is applied, a degree of heat is given to the wick, sufficient to destroy the affinity of its constituent parts; part of the tallow is instantly melted, volatilised, and decomposed; its hydrogen takes fire and the candle burns.

768. *Why does the light of the candle improve after a time?*

Because, as the heat increases, the amount of gas evolved from the candle is increased, the flame enlarges, and the *carbon* taking part in the combustion improves the light of the flame.

769. *Why does the tallow ascend the wick?*

It is liquified by the heat of the flame, and is drawn up the wick by *capillary attraction*; when it comes into contact with the flame, it is converted into inflammable gas.

770. *Why does the wick of the candle turn black?*

Because the charcoal of the cotton becomes predominant, when its other constituents are driven off. The flame surrounding it, it cannot undergo the *oxidating* influence of the air; it therefore remains for a considerable time in its natural state.

771. *Why do tallow candles require snuffing?*

Because the wick forms a support for an *accumulation of soot*, which is produced by imperfect combustion.

772. *Why, if the candle be made of wax, does it not require snuffing.*

Because the thinness of the wick permits it to bend on one side, when its length is too great for its verrtical position; its extremity comes then into contact with the air, and is completely burnt or decomposed. This small wick therefore performs the office of *snuffing itself.*

773. *Why may a thinner wick be used for a wax candle than a tallow one?*

Because wax is less fusible than tallow, therefore it *affords a better support* to the wick.

774. *Why does a pale envelope of light* (A) *surround the flame of a candle?*

Because the liberated carbon, coming in contact with the oxygen of the air, becomes at once *oxidised,* and passes away as an *invisible gas.*

775. *Why may the pale fringe* (A) *be rendered more visible by shutting off from the eye the luminous part of the flame by the hand?*

Because it prevents the *flow of intense light* to the eye, which renders the pale envelope of the flame invisible.

"If then God so clothe the grass which is to-day in the field, and to-morrow is cast into the oven, how much more will He clothe you, O ye of little faith."—LUKE XII.

776. *Why is there a luminous flame* (B) *within the pale fringe?*

Because there the *carbon* and *hydrogen*, in an intensely heated state, combines with oxygen, and burn with a bright yellow light.

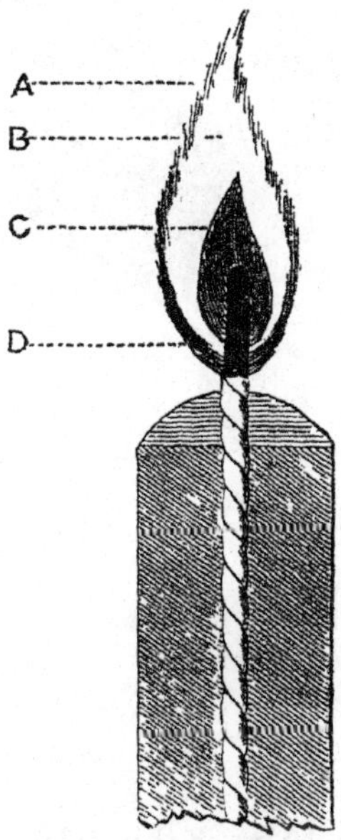

DIAGRAM SHOWING THE COMBUSTION OF A CANDLE.

777. *Why does the centre of the flame* (C) *appear like a dark cone?*

Because there the gases just formed by the decomposition of the tallow are escaping, but *do not ignite* until they arrive at the second, or luminous cone.

778. *Why has the lower portion of the flame a blue colour?*

Because the hydrogen having a very strong tendency to combine with oxygen, some of it *at once ignites.* Hydrogen, without carbon, burns with a blue light, which accounts for the colour of the lower part of the flame.

779. *Why does the light of a candle diminish as the snuff increases?*

Because the snuff radiates the heat of the flame, and *reduces its temperature.* Everything that tends to lower the temperature of flame diminishes the intensity of its light.

780. *Why do currents of cold air* (B) *flow towards the flame of a candle?*

Because as the air is rendered hot by the flame, and mixed with the gas produced by combustion it rises upward in *heated* currents. The colder and denser air therefore *rushes towards the flame,* to supply the place of the warmer air that has passed away.

"Every tree that bringeth not forth good fruit is hewn down and cast into the fire."—MATTHEW VII.

781. *Why do the currents of air* (A) *ascend?*

Because, being expanded by heat, they are rendered *specifically lighter* than the surrounding air.

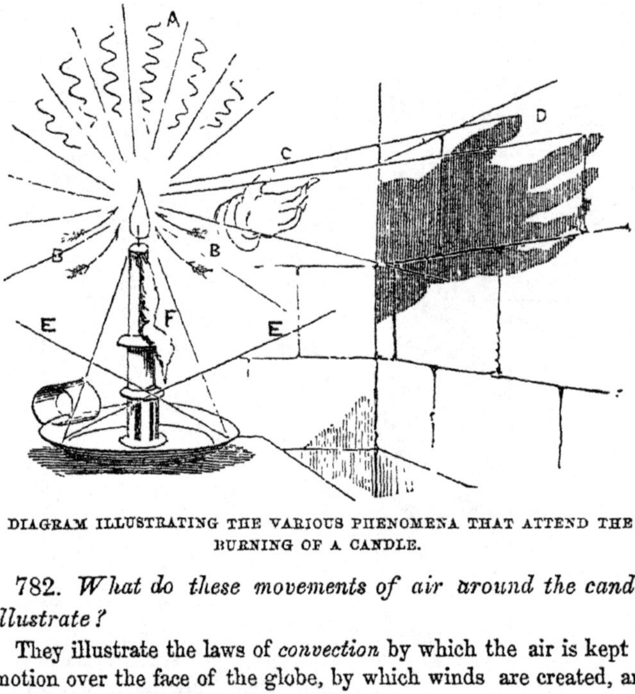

DIAGRAM ILLUSTRATING THE VARIOUS PHENOMENA THAT ATTEND THE BURNING OF A CANDLE.

782. *What do these movements of air around the candle illustrate?*

They illustrate the laws of *convection* by which the air is kept in motion over the face of the globe, by which winds are created, and in consequence of which storms arise.

783. *Why do rays of light proceed in every direction from a candle flame?*

Because light *radiates* in all directions from luminous bodies.

784. *Why do shadows of objects interposed between a light and a wall appear very large.*

Because light travels only *in straight lines.* The hand in the diagram intercepts all the rays of light that fall upon it, and

"Out of his mouth go burning flames, and sparks of fire leap out."—
JOB XLI.

allows only those to pass that are diverging from flame at a considerable angle. Hence the light is turned off from the wall over a space much larger than the hand, causing *a shadow of considerable magnitude.*

785. *Why are rays of light, when they fall upon a polished candlestick, reflected?*

Because bright surfaces, and especially those of metal, *reflect the rays of light,* and turn them back at an angle equal to the angle of incidence, E E, (*See* 61.)

786. *Enumerate the various phenomena illustrated by the burning of a candle, as represented in the diagram?*

The burning of a candle illustrates the *radiation* of light and heat; the *reflection* of light; the distribution of heat by *convection,* and the cause of *winds;* the radiation of light only in *straight lines,* causing *magnified shadows* of objects; the effect of the *caloric of fluidity,* in rendering solid bodies fluid, &c. &c.

CHAPTER XXXVII.

787. *How does atmospheric air support life?*

It supports life in the same manner that it *supports flame,* by giving out its *oxygen* and *caloric* to the blood.

788. *What is given off with the breath from our lungs?*

Carbonic acid gas, formed by the carbon of the blood combining with the oxygen of the air, and *water.* The *nitrogen* of the air, from which the oxygen has been separated, is thrown back with our expired breaths.

789. *What effect has air upon the animal system when repeatedly breathed?*

It produces disease, and ultimately death by suffocation.

790. Two men were employed to visit a wreck in a diving bell, in the Bay of Dublin. The machinery by which they were to signal their wants to people attending upon them became out of order. The supply of air was insufficient, and as they could not convey signals through the derangement of the machinery they died. When the bell was raised they were both found to be quite dead. They were not *drowned,* but died like the unhappy people in the Black Hole at Calcutta, for want of a supply of pure air.

"Out of his nostrils goeth smoke, as out of a seething pot or cauldron."—
JOB XLI.

791. *How do clothes conduce to preserve the heat of our bodies?*

Clothes keep our bodies warm by preventing the *sudden escape of heat* from the surface of our bodies.

792. *Why do clothes prevent the heat of our bodies from escaping?*

They keep our bodies warm, in consequence of the air which they enfold within them; all confined bodies of atmosphere being *non-conductors* of heat.

793. *Why do light spongy substances, such as furs and down, afford the warmest clothing?*

Because they contain air in the spaces of their texture, which forms a *non-conducting shield* around the body.

794. *What is cold?*

Cold is simply the *absence of heat.* There is no element to which the effects of cold can be attributed.

795. *Why does a piece of glass held before a fire keep off the heat, but transmit the light?*

Because glass is *less transparent* to *heat* than it is to *light.* It therefore separates the heat from the light, allowing the one to go through it, but detaining the other.

796. *Why does the heat of the sun pass through a glass window, while the heat of the fire will not so readily pass through a pane of glass?*

The cause is unknown; but it is conjectured that in the sun's rays the heat and light are *more intimately blended* than they are in the rays of a fire.

797. *Why does the heat of a hot body always endeavour to escape and pass to colder bodies?*

Because caloric always seeks to diffuse itself. For this reason all the furniture in a room, the tables, the bureaus, the carpets, and the walls, all become heated alike.

"Thus saith the Lord, Let not the wise man glory in his wisdom, neither let the mighty man glory in his might, let not the rich man glory in his riches."—JEREMIAH IX.

798. *If you put a stick of wood with one end in the fire, why can you let it burn off without feeling the heat at the other end?*

Because wood is a *bad conductor* of heat. It will not transmit heat as the poker does. Therefore the handle of the poker becomes hot; but the end of the wood remains cool.

799. *Why does straw tied around pumps in winter prevent the water from freezing?*

Because every stalk of straw is hollow, and *full of air;* it is the air confined in the stalks of the straw that makes it so good a protector of many things from cold.

800. *Why does burning wood snap and throw out sparks?*

Because the air imprisoned in the wood expands by heat and *bursts the cells* where it is confined, carrying very small bits of burning wood with it.

801. *Why do wheelrights heat the iron tire of a wheel before they put it on?*

Because the tire *when heated is larger than when cold*, in consequence of the expansion of the metal. As it cools it shrinks, and binds the wheel more firmly together.

802. *Why do they pour cold water upon the tire directly it is fixed?*

Because the cold water *takes away the heat of the iron,* and causes it to cool and shrink.

803. *Why does boiling water make a simmering noise?*

The noise is made by little bubbles of steam which are formed at the bottom of the kettle. These *rush upwards,* and being attracted to the *sides of the kettle,* they make a commotion which sets the metal in vibration, and the kettle "sings."

804. *Why does steam put steam-engines in motion?*

It is the tendency of steam to *expand* which makes it so very powerful?

When steam rushes out of a *small valve* in a steam-engine it instantly spreads out into a *large cloud of vapour* which fills the air for some distance.

Within the engine steam is *prevented from expanding*, and is allowed to escape only through certain openings. It is in *forcing its way through these openings* that it moves the machinery, and sets the wheel in motion.

805. *Why does the lid of a tea-kettle rattle?*

Because the bubbles of steam as they rise *force up the lid*, just as the steam of a locomotive forces open its valves. When the lid rises, some of the steam rushes out; this diminishes the pressure on the inside, and the lid falls again. This rising up and down makes a rattling noise?

806. *Why are safety valves attached to steam-engines?*

For the purpose of letting off the steam, before it presses with *sufficient force to burst the boiler*. The valve is made to open by the force of the steam, at a *less pressure* than that which would rend the metal of the boiler. Therefore when the valve opens, the pressure of the steam is kept down.

807. *Why do boilers sometimes burst?*

Because sometimes there happens to be a bad place in the metal which will not bear *so much pressure as the valve*. The boiler therefore gives way in the weak part, and steam rushes out with mighty force.

808. *How much more space does steam occupy than water?*

A *pint of water*, converted into steam, occupies as much space as *seventeen hundred pints* of water. It is in seeking to expand to this greatly increased space that steam displays such mighty force.

809. In other cases the boilers are carelessly allowed to get empty, and the fire makes them *very hot*. Then, when cold water is suddenly let into them, a great deal of steam is *instantly generated*. And the sudden addition of this force of steam, being greater than the valve can keep in check, rends the metal.

The boilers in *kitchen ranges* are frequently cracked by the same cause.

"The father of the righteous shall greatly rejoice; and he that begetteth a wise child shall have joy of him."—PROVERBS XXIII.

CHAPTER XXXVIII.

810. *Why does a humming-top make a humming noise?*

Because the hollow wood of the top vibrates, and the edges of the hole in its sides *strike against the air as it spins;* the air is thereby set in vibration.

811. *Why does a peg-top hum less than a humming-top?*

Because, *being a solid body of wood,* and having no *hole in its sides,* its particles are *not so easily thrown into vibration;* consequently it does not so readily impart vibrations to the air.

812. *Why does a peg-top sometimes hum, and at other times not?*

Because, if it is spun with *great force,* and its peg is *struck sharply* against the pavement, *the wood is set in vibration,* and the surface of the top, repelling the air by its rapid motion, causes *vibratory waves.* But if it be spun with insufficient force, *the wood is not set in vibration.*

Fig. 23.—HUMMING-TOP BEFORE SPINNING.

Fig. 24.—HUMMING-TOP SPINNING.

813. *Why do we see the figures painted upon the humming-top, before it spins, but not while it is spinning?*

Because the rapid whirling of the top brings the images of its different parts so quickly in succession upon *the retina of the eye,*

"Train up a child in the way he should go; and when he is old, he will not depart from it."—PROVERBS XXII.

that they *deface each other*, and *impart an impression of coloured rings, instead of definite objects.*

814. *Why does a top stand erect when it spins, but fall when it stops?*

Because the top is under the influence of, and is balanced between *opposing forces.* The rapid rotation of the top gives to all its particles a tendency to *fly from the centre.* If the atoms of the wood were not held together by the *attraction of cohesion,* they would fly away in a circle outward from the top, *just as drops of water fly off from a mop, while it is being twirled.* If you take a spoonful of sand, salt, or dust, and drop it upon the top, it will be scattered in a circle, just as the atoms of the top would be, *if they were free to separate,* but not with the same force, because the atoms of the salt, &c., not being in an active state of rotation, would only be influenced *by momentary contact with the rotating body.* This tendency of the particles of a rotating body to fly outward from the centre, is called *the centrifugal force.*

Centrifugal.—From two Latin words meaning receding from the centre.

The other force influencing the top is *the attraction of gravitation:* the attraction which, were the top not spinning, would draw it towards the earth. The "spill" projecting from the bottom of the top *stands in the line in which the top is drawn towards the earth* and keeps it from obeying the law of gravitation. Therefore the rotatory motion given to the top, by the rapid unwinding of the string, and the tendency of its atoms to fly outward, *balance the top* upon the line in which it is drawn to the earth, and which is occupied by the spill, which prevents it falling to the ground.

815. *Why does a top first reel around upon the spill, then become upright, and "sleep," and then reel again, and fall?*

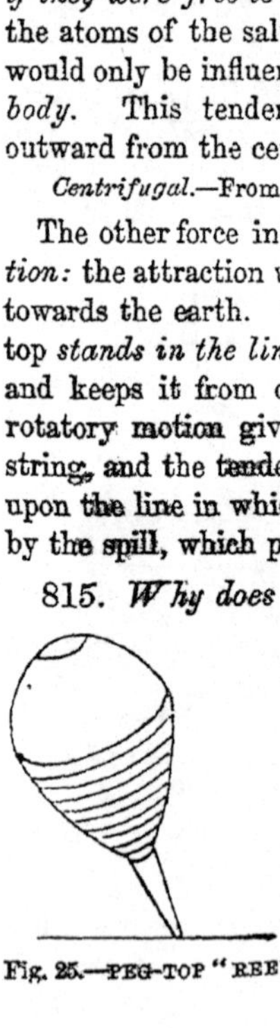

Fig. 25.—PEG-TOP "REELING."

Because, in being thrown from the hand, the top is delivered a little out of the perpendicular, but the spill *is rounded off at the point,* and when the top is rotating rapidly, the gravitative force which attracts the top to the ground continually acting upon it, *draws the weight of the top on to the extreme centre of the round point.* When the rotation subsides, and the centrifugal force is weakened, then the top *is no longer balanced upon the extreme point of the spill,* but falls upon it sides, until the force of gravitation is exerted *beyond the line of the spill,* upon the body of the top, and then it falls to the ground.

816. *Why does a top "sleep?"*

Because at that period of its spinning, which is called "sleeping," the *centrifugal* and the *gravitative forces* acting upon the top, are *nearly balanced;* and the top, obeying chiefly the *rotatory force,* appears to be in a state of comparative rest.

817. *Why does the top cease to spin?*

Because *the friction of the air against its sides,* and the *friction of the spill against the ground,* act in opposition to the *rotatory force,* which is a temporary impulse applied by external means—the hand of the person who spins it—and as soon as this *applied force* is expended, the top yields to the law of gravitation, which is *a permanent and ever-prevailing force.*

818. *Why does a marble revolve, as it is propelled along the ground?*

Because, in propelling the marble, *the thumb impels the upper surface forward, and the finger draws the under surface backward.* This gives a tendency to the upper and lower hemispheres of the marble *to separate,* which they would do, but for the *cohesion of the atoms* of the marble. The upper part of the marble, therefore, rolls forward, *drawing after it the under part,* which acquires a forward motion by the force with which it is drawn upward, and in this way the opposite portions of the marble act upon each other in the successive revolutions.

When the marble strikes upon the earth, a new influence is exerted upon it, which is *the friction of the earth* upon the surface that comes in contact with it; but the upper part of the marble, being free, *overcomes the friction acting upon the lower part,* and thus the marble continues to progress, until *the applied force which projected it is expended.*

"Better is a poor and a wise child, than an old and foolish king who will no more be admonished."—ECCLESIASTES IV.

819. *Why does a striped marble appear to have a greater number of stripes when rolling, than when at rest?*

Because the stripes are presented in *rapid succession* to the eye; and as the eye receives *fresh impressions of stripes before the previous impressions have passed away*, the stripes appear multiplied.

Fig. 26.—MARBLE AT REST

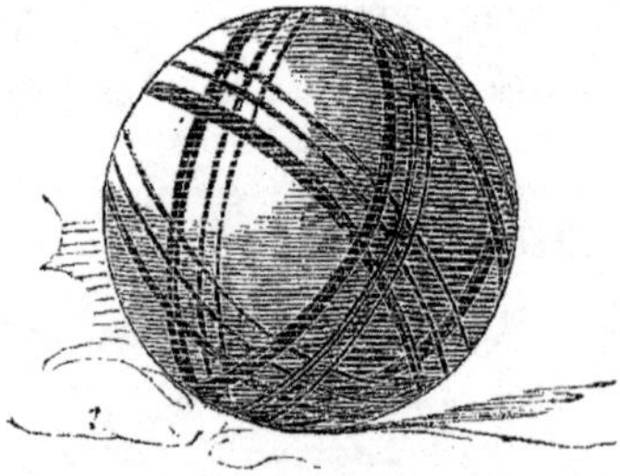

Fig. 27.—MARBLE ROLLING.

820. *Why does a marble rebound when dropped upon the pavement?*

Because the force of its fall to the earth *compresses the atoms* of which the marble is composed; and the atoms then exert the force of *elasticity to restore themselves to their former condition;* and by the exercise of this force the marble is *repelled*, or *thrown upward from the pavement.* Although a marble may be made of very hard stone, yet that stone may be *elastic*, and possess, though in a much less degree, *the same kind of elasticity which causes the India-rubber ball to rebound from the earth.*

821. *Why does a marble, assuming it to be impelled with equal force, roll further on ice than on pavement, and further on pavement than on a pebble walk?*

Because the *friction* is greater upon pavement than upon ice, and greater upon a pebble walk than upon pavement.

822. *How many forces contribute to stay the progress of a rolling marble?*

The friction of the *air*, the friction of the *earth*, and the *attraction of gravitation*, which tends to bring all bodies to a state of rest.

823. *Why do the stripes upon a marble disappear when it is spun with great velocity?*

Fig. 28.—MARBLE SPINNING RAPIDLY.

Because, as in the case of the humming-top, the different parts of the surface are *brought so rapidly in succession to the sight*, that they *deface or confuse* the impressions upon the retina.

824. *Why are rings most perceptible at the opposite points, or poles, of the marble?*

Because the point, or pole, *upon which the marble spins*, and that which *corresponds to it*, on the upper surface, travel *less rapidly* than the central portions, which being of a larger circumference, pass through a greater amount of space, in the same period of time. The stripes at the *poles* of the marble, are, therefore visible, while those at its *equator* are imperceptible. (*See* 522.)

CHAPTER XXXIX.

825. *Why are soap-bubbles round?*

Because they are *equally pressed upon all parts of their surface* by the atmosphere.

826. *Why are bubbles elongated when being blown?*

Because the *unequal pressure of the current of breath* by which they are being filled, alters the *relative pressure* upon the outer surfaces.

827. *Why does the bubble close, and become a perfect sphere, when shaken from the pipe?*

"Children's children are the crown of old men; and the glory of children are their fathers."—PROVERBS XVII.

Because the *attraction of cohesion* draws the particles of soap together, directly the bubble is set free from the bowl.

Fig. 29.—BLOWING SOAP BUBBLES.

828. *Why do bubbles, blown in the sunshine, change their colours?*

Because the films of the bubbles constantly change in thickness, through the atoms from the upper part descending towards the bottom, and therefore the varying thickness of film *refracts, in different degrees, the rays of light.*

829. *Why do bubbles burst?*

Because the atoms that compose their films *fall towards the earth by gravitation;* the upper portion of the bubbles then *becomes very thin,* and as the denser air of the atmosphere *presses towards the warm breath within the bubble, it bursts the film.*

See 236. 237, *etc.*, 501, *etc.*

830. *Why do balloons ascend in air?*

Because the air or gas which they contain is *specifically lighter than the atmosphere; the atmosphere, therefore, forces itself underneath the balloon,* by its own tendency towards the earth, and the balloon is thereby raised upwards. *A balloon is but a larger kind of bubble, made of stronger materials.*

831. *Why does an air-balloon become inflated when the spirit set upon the sponge is lit?*

"A wise son heareth his father's instruction."—PROVERBS XIII.

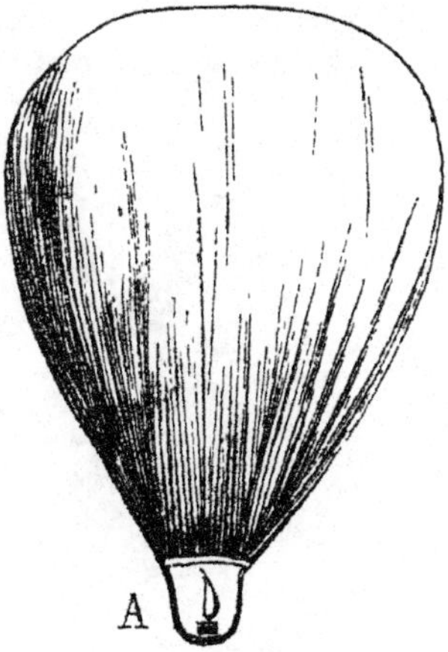

Fig. 30. AIR-BALLOON.

Because the *heat* of the flame, and the *burning of the spirit*, A, create a volume of *rarefied*, or *thin air*, which inflates the balloon, and makes it *specifically lighter* than the surrounding medium.

832. *Why do balloons sometimes burst when they ascend very high?*

Because, as they get into the *thinner air*, which exists at *high altitudes*, the gas within them expands, and the coating of the balloon is burst asunder.

833. *Why does the gas of balloons expand in thin air?*

Because the air exerts a *less amount of pressure* upon the air or gas contained in the balloons.

834. *Why do parachutes fall very gradually to the ground?*

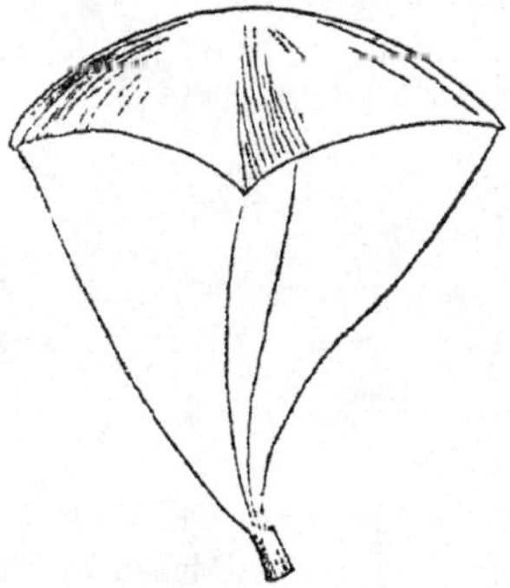
Fig. 31.—PAPER PARACHUTE.

Because the *air*, coming in contact with the *under surface* of the expanded head of the parachute resists its downward progress.

835. *Why does a shuttlecock travel slowly through the air?*

Because the air acts upon the feathers of the shuttlecock, in the same manner as it does upon the parachute—it strikes against their expanded surface, and resists their progress through the air.

836. *Why does the shuttlecock spin in the air?*

"Come ye children, hearken unto me, I will teach you the fear of the Lord."—PSALM XXXV.

Because the surfaces of the feathers fall upon the air *obliquely*, or slantingly, and therefore, as the shuttlecock descends, it turns in the air.

Fig. 32.—BATTLEDORE AND SHUTTLECOCK.

837. *Why do we hear a noise when we strike the shuttlecock with the battledore?*

Because the *percussion* of the shuttlecock upon the parchment of the battledore causes it to vibrate, and the vibrations are imparted to the air.

838. *Why is the sound a dull and short one?*

Because the vibrations of the parchment are *not very rapid*, therefore there is *little intensity* in the vibrations of the air.

839. *Why does the exercise, afforded by playing battledore and shuttlecock, make us feel warm?*

Because it makes us breathe *more freely*, and causes the *blood to flow faster;* we, therefore, inhale more *oxygen*, which produces heat by combining with the *carbon* of our *blood.*

840. *Why does a kite rise in the air?*

A kite rises in the air by the force of the wind, which *strikes obliquely* upon its *under surface.* The string is attached to the "belly-band" in such a manner that it is nearer the *top* than the *bottom* of the band: this causes the bottom of the kite, when its surface is met by the wind, to recede in the direction of the wind;

the top is accordingly *thrown forward*, and the kite is made to *lie obliquely* upon the current of air moving against it. The kite then being *drawn by the string in one direction*, and *pressed by the air in another direction*, moves in a line which *describes a medium between the two forces acting upon it.*

841. *Why does the kite-string feel hot when running through the hand?*

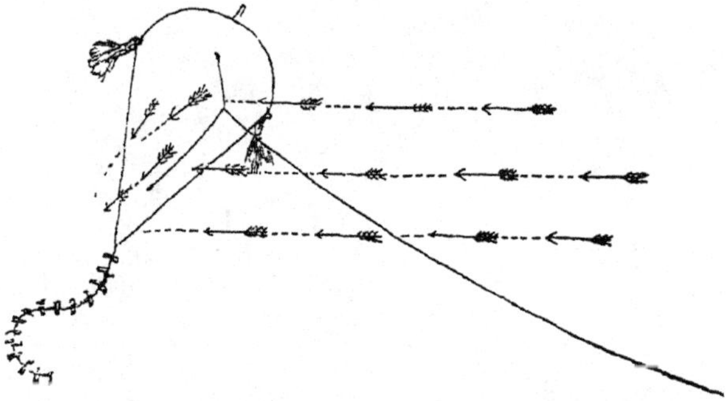

Fig. 33.—DIAGRAM EXPLAINING THE FLIGHT OF A KITE.

Because the *rapid friction* sets free the *latent heat* of the *string*, attracts the heat of the *hand* to the spot where the friction occurs, and sets free the latent heat of the *air*, which follows the *string* through the hand, and is compressed by the friction.

842. *Why does running with the kite cause it to rise higher?*

Because it *increases the force* with which the wind strikes upon the surface of the kite. If a person were to *run with a kite at the rate of five miles an hour, through a still air*, the effect would be *equal to a wind flying at the rate of five miles an hour* against a kite held by a *stationary string*.

843. *Why does the flying-top rise in the air?*

Because its wings *meet the air obliquely*, just as the surface of the kite does. And the *twirling of the top*, causing the oblique surfaces of its wings to strike the air, produces *the equivalent effect of a wind from the earth blowing the top upwards.*

844. *Why does the flying-top return to the earth when its rotations are expended?*

Because the *reaction* produced by its wings striking upon the air, is insufficient to counteract the *attraction of gravitation.*

Fig. 34.—FLYING-TOP.

Fig 35.—PEA AND PIPE.

845. *Why does a pea, into which a pin has been stuck, dance in suspension upon a jet of air blown through a pipe?*

Because the jet of air, being *slightly compressed* under the *convex* form of the pea, by the weight of the pin, forms a *concave cup of air*, in which the pea rests.

In the case put, it is supposed that the pin is *passed through the pea* until its head comes in contact with it. The pin is dropped into the hole of the pipe, and the breath is then applied, the pipe being held upright. The pea will rise in the air, and be suspended upon the jet, while the point of the pin will rotate around the stem of the pipe. There are other methods of fixing the pin which alter the result, and require a different explanation to that given above.

LESSON XL.

846. *Why does a mouse, painted upon one side of a card, and a trap upon the other, represent to the eye a*

mouse in a trap when the card is rapidly twirled upon a string?

Because the image of the mouse is brought to the retina of the eye before the image of the trap has passed away. The two impressions, therefore, *unite upon the retina*, and produce the image of a mouse in a trap.

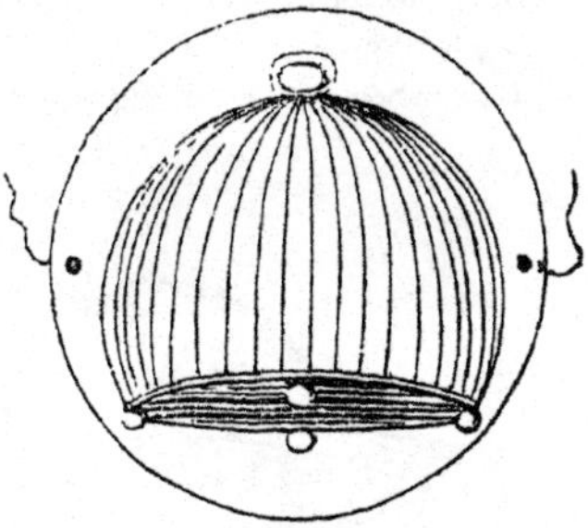

Fig. 36.—CARD WITH MOUSE-TRAP.

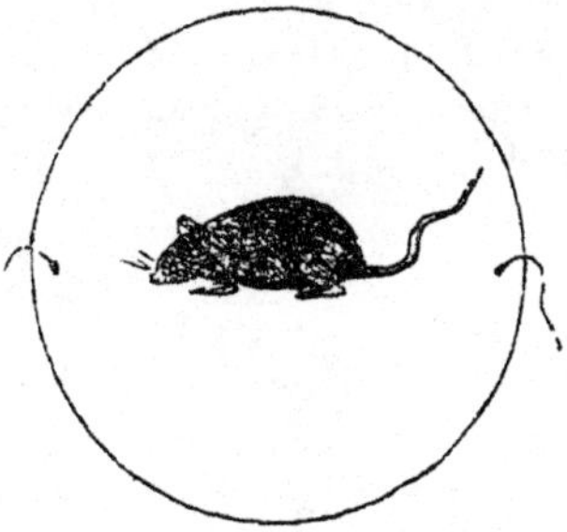

Fig. 37.—REVERSE OF CARD WITH MOUSE.

847. *Why will a bow stretched out of its natural position, propel an arrow through the air?*

Because its substance, being *highly elastic*, the particles thereof seek to restore themselves to their former state, as soon as the resisting power is withdrawn. The *force* derived from this elasticity, is communicated to the arrow by the string against which it is placed.

848. *Why is the arrow propelled forward?*

Because the elasticity of the bow, *acting equally upon its two ends*, to which the string is fastened, produce a line of force in a *diagonal direction.* It thus illustrates the law, that *when a body is acted upon by two forces at the same time, whose directions are inclined to each other, it will not follow either of them, but will describe a line between the two.*

849. *What forces tend to arrest the flight of the arrow?*

The *friction of the air*, and the *attraction of gravitation.*

"My son, give, I pray thee, glory to the Lord God of Israel, and make confession unto him."—JOSHUA VII.

850. *Why are feathers usually fastened to the ends of arrows?*

Because the *greater friction* of air acting upon them, opposes the progress of that part of the arrow in a greater degree than it does the other portion. The effect is, *to keep the point of the arrow forward,* and in a straight line with its opposite extremity. If the arrow were shot the reverse way from the bow, it would *turn round,* in the course of its flight, in consequence of the friction of the air, offering greater resistance to the progress of the feathered end.

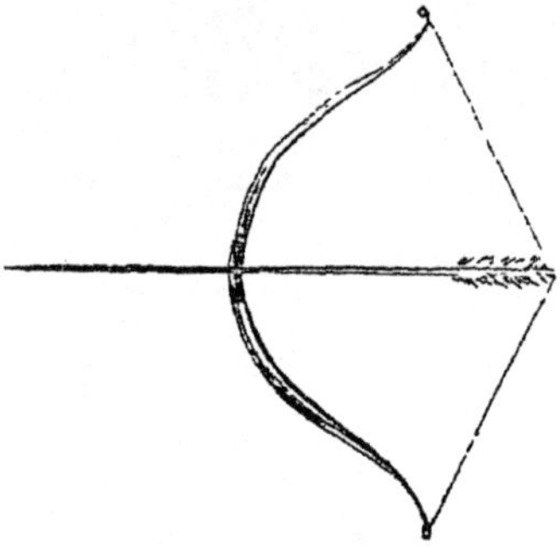

Fig. 38.—BOW AND ARROW.

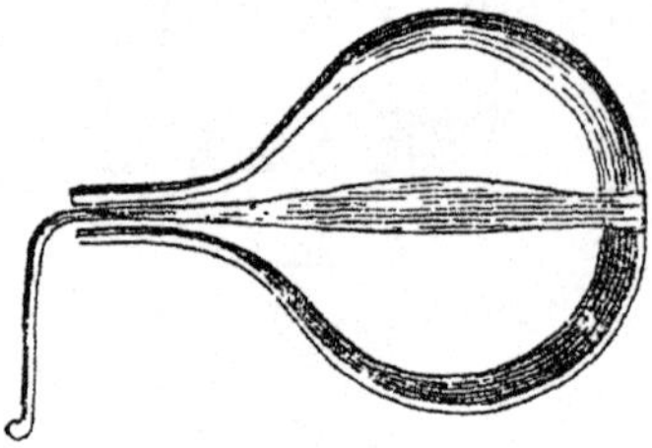

Fig. 39.—JEW'S HARP.

851. *Why does a Jew's harp give musical sounds?*

Because the *vibrations of the metal tongue* are communicated to the ear.

852. *Why will not the Jew's harp produce loud sounds unless it is applied to the mouth?*

Because the vibrations are not very intense, but when it is blown upon by the breath, the air is pressed upon it, and the vibrations are thereby rendered more powerful.

853. *Why does the alteration of the arrangement of the mouth, affect the formation of the sounds?*

Because it sends the air to the tongue of the harp in *a greater or lesser degree of compression.*

"Hear, ye children, the instruction of a father, and attend to know understanding."—PROVERBS IV.

854. *Why does the pressure applied to the handle of an air pistol propel the cork?*

Because, between the cork A and the air-tight piston C, there is a *closed chamber of air* B. When the handle D, which moves the piston C, is rapidly pushed in, it *compresses the air* until it is so much condensed, that it forces out the cork A.

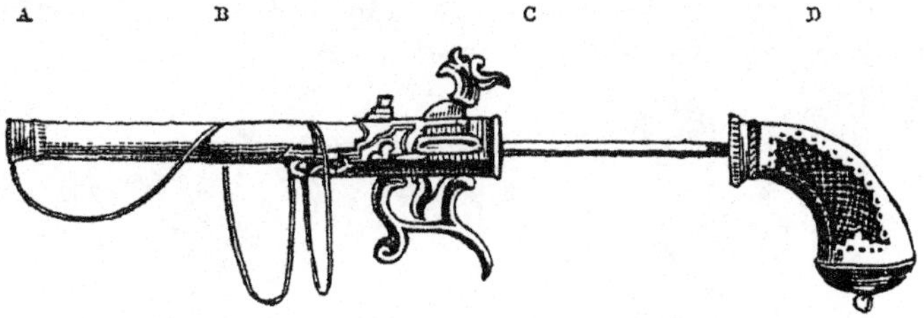

Fig. 40.—AIR PISTOL, OR "POP-GUN."

855. *Why must the handle be drawn out, before the cork is placed in?*

Because otherwise a partial *vacuum* would be formed between A and C, and there would not be sufficient air to force out the cork by the return of the piston C D.

856. *Why does water rise in a syringe when the handle is drawn out?*

Because the pressure of the air on the water outside of the syringe, forces it into the space vacated by the drawing up of the handle, and where, otherwise, a *vacuum* would be formed.

Fig 41.—SYRINGE, WITH JET OF WATER.

857. *Why does not the water run out when the syringe is raised?*

Because the pressure of the air upon the small orifice resists the weight of the water.

858. *Why does the water leak out, but not run?*

Because water has a tendency always to *move to the lowest point,* but as the air does not enter freely the water cannot escape. It therefore *drops*, as small portions of the air enter.

859. *Why cannot the handle be pressed in, if the finger is applied to the orifice?*

Because water is not *compressible*, like air; it must therefore escape before the handle can be pressed in. Air may be forced into a much smaller compass than is natural to it; but it is impossible to *compress water* in any great degree.

Fig. 42.—"SUCKER."

Fig. 43.—HOOP.

860. *Why does a "sucker" raise a stone?*

Because underneath the sucker *a vacuum* is formed and the external air, pressing on all sides *against the vacuum*, lifts the stone. The term "sucker" is founded upon the mistaken notion that the leather "sucks," or "draws" the stone. That such is not the case is evident: if, when the stone is suspended, a pin's point be passed under the leather, so as to open a small passage for the air, the stone will *drop instantly*.

861. *Why does a hoop roll, without falling to the ground?*

Because the *centrifugal force* gives it a motion which is called the *tangent to a circle*—that is, a tendency in all its parts *to fly off in a straight line*. When a piece of clay adhering to the hoop flies off, it leaves the hoop in a line which is straight with the part of the surface from which it was propelled; this line is *the tangent to the circle of the hoop;* and the tendency of all the parts of the hoop to fly off in this manner, counteracts the attraction of the earth, so long as the hoop is kept in motion.

862. *Why does the hoop, in falling, make several side revolutions?*

Because its onward movement, not being quite expended, influences the *centre of gravity of the hoop,* and changes its line of direction. The hoop is also elastic, and when its sides strike the earth, they spring up again, and continue turning until the opposing forces are overcome by the *attraction of gravitation.*

863. *Why will a little boy balance a large boy on a see-saw?*

Because the "see-saw" may be placed so that its ends are at *unequal distances from the centre.* This gives the little boy the power of *leverage,* by which is meant the increase of power, or weight, by *mechanical means.*

Fig. 44.—BOYS AND "SEE-SAW."

864. *Why does the little boy sink to the ground when the larger boy slightly kicks the earth?*

Because the larger boy, by kicking against the earth, opposes by mechanical force the *attraction of gravitation* acting upon him, and he becomes *temporarily* less attracted to the earth than the little boy.

865. *Why can the little boy, if he choose, keep the big boy up, when once he is up?*

Because, as the big boy is then on *an inclined plane* with the *fulcrum*, or centre upon which the see-saw moves, the arm of *the lever*, upon which the big boy sits, is *relatively shortened*, and he has then *less mechanical power*. Also, a portion of the weight of the larger boy is transmitted along the lever *to the arm upon which the little boy sits.*

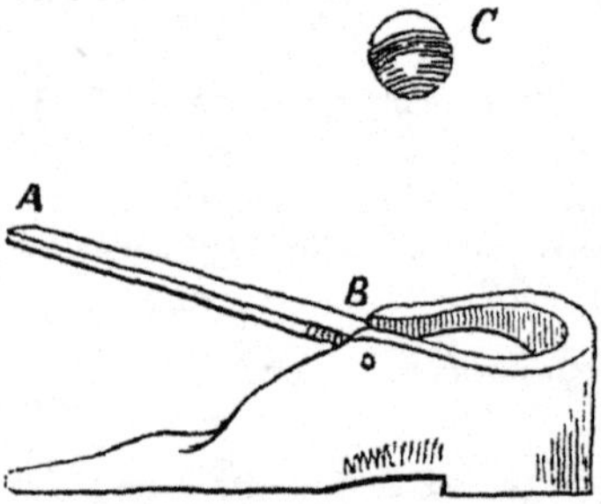

Fig. 45.—TRAP AND BALL.

866. *Why is the ball propelled upward, in the game of trap and ball, when the trigger is struck?*

Because, when the trigger is struck at A, it is forced down wards, turning upon the fulcrum B, the opposite end, forming the spoon, is thereby forced upwards, describing a small *arc*, or curved line; but directly the ball is set free from the spoon, it rises in a *right line* with the *direction it was taking, at the moment it was set free.*

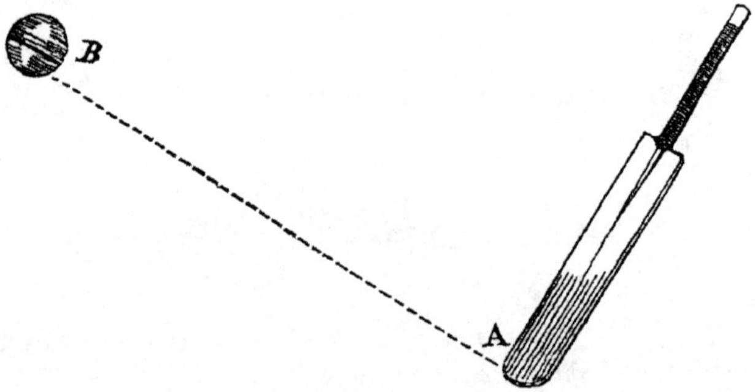

Fig. 46.—BAT AND BALL.

867. *What principles of natural philosophy are illustrated by the results of bat and ball?*

Percussion, when the bat strikes the ball; *rotatory motion,* when the ball is sent whirling away; *momentum,* which it acquires by velocity; *elasticity,* when it rebounds from an object against which it strikes; *reflected motion,* when it is turned by a body upon which it impinges; *friction,* as it rolls along the ground; the *communication of force,* when it sets another body in motion against which it strikes; *gravitation,* when it falls to the earth; and *inertia,* when it lies in a state of rest.

868. *Why do pith-tumblers always pitch upon one end?*

Fig. 47.
PITH-TUMBLER.

Because the *lead* B is *specifically heavier* than the *pith* to which it is attached; it therefore always falls undermost; and as the lead is rounded off, just like the spill of a top, after the head has oscillated a little, and expended the force of the momentum of its fall, it will settle upon its *centre of gravity,* or the point through which it is *attracted to the earth.*

869. *Why do the figures upon the "Thaumatrope" appear to dance, when they are made to revolve before a mirror?*

Because the eye, in looking through the holes in the card, towards the reflections in the mirror, receives *a rapid succession of impressions.* As the figures upon the card are represented in a graduated series of positions—the *first* one standing upright, the *second* with his knees a little bent, the *third* a little more bent, as in the act of springing, and so on, the *figure* being in each case *the same,* but the position *slightly altered,* imparts an impression to the mind, through the eye, that *one figure* is passing through a *series of motions.*

Thaumatrope.—From two Greek words, meaning *wonder* and *to turn.*

We have said enough, we hope, to show that even the play-hours of children may be made instructive to them; and that the simplest toys may be used to illustrate some of the grandest laws of nature. Nor may this kind of instruction be confined to children alone. Grown-up people, whether participators in the sports of youth, or simple observers of their games, may gain instruction for themselves, and be the better teachers of their children, by taking an interest in their enjoyments, and giving to their minds, through the attractiveness of pastime, a taste for observing and estimating the varied phenomena which present themselves.

Moreover, we think that parental government acquires a greater power when

"Jesus said, Suffer little children, and forbid them not, to come unto me; for of such is the kingdom of heaven."—MATTHEW XIX.

it leans towards the natural desires of childhood, and wins those desires into a proper direction. Love existing between parent and child is the best tie to home, and the strongest incentive to duty. There is also something in the gentleness of childish nature which may influence for good the sterner mould of man, too often warped and clouded by the cares of life.

Fig. 48.—THAUMATROPE, OR "WONDER-TURNER."

In Kay's "Life of Sir John Malcolm," we find an admirable and apt passage. Sir John says:—"I have been employed these last few hours with John Elliot, and other boys, in trying how long we could keep up two cricket-balls. Lord Minto caught us. He says he must send me on a commission to some very young monarch, for that I shall never have the gravity of an ambassador for a prince turned of twelve. He, however, added the well-known and admirable story of Henry IV. of France, who, when caught on all fours carrying one of his children, by the Spanish envoy, looked up and said, 'Is your excellency married?' 'I am, and have a family,' was the reply. 'Well, then,' said the monarch, 'I am satisfied, and shall take another turn round the room,' and off he galloped, with his son on his back flogging and spurring him. I have sometimes thought of breaking myself of what are termed boyish habits; but reflection has satisfied me that it would be very foolish, and that I should esteem it a blessing that I can find amusement in everything, from tossing a cricket-ball, to negotiating a treaty with the Emperor of China. Men who will give themselves entirely to business, and despise (which is the term) trifles, are very able, in their general conception of the great outlines of a plan, but they feel a want of knowledge, which is only to be gained by mixing with all classes in the world, when they come to those lesser points upon which its successful execution may depend."

"Whether therefore ye eat, or drink, or whatsoever ye do, do all to the glory of God."—CORINTH. X.

CHAPTER XLI.

*869. *Why do we eat food?*

Because the atoms of which our bodies are composed are *continually changing*. Those atoms that have fulfilled the purposes of nature are removed from the system, and, therefore, new matter must be introduced to supply their place.

870. *Why do we eat animal and vegetable food?*

Because their substances are composed of *oxygen*, *hydrogen*, *carbon*, and *nitrogen*—the four chemical elements of which the human system is formed. They are, therefore, capable of nourishing the body, after undergoing digestion.

871. *Why do we masticate our food?*

Because mastication is *the first process towards the digestion of food*. Before animal or vegetable substances can nourish us, their condition must be entirely changed, their *organic* states must be dissolved, and they must become simple matter, in a homogeneous mass, consisting of the four chemical elements necessary to nutrition, and they must again be restored to an organic condition.

872. *Why does saliva enter the mouth when we are eating?*

Because, in addition to the *mechanical* grinding of the food by the action of the teeth, it is necessary that it should undergo certain chemical modifications to adapt it to our use. There are placed, therefore, in various parts of the body, *glands*, which secrete peculiar fluids, that have a chemical influence upon the food.

The first of these glands are the *salivary glands of the mouth*, which pour out a clear watery fluid upon the food we eat, and which fluid has been found to possess a property which contributes to the digestion of food.

The moisture afforded by the salivary secretion is also necessary to enable us to swallow the food.

873. *Why does the salivary juice enter the mouth just at the moment that we are eating?*

"And the Lord said unto him, Who hath made man's mouth? or who maketh the dumb, or the seeing, or the blind? have not I the Lord?"—EXODUS IV.

Because the glands, which are buried in the muscles of the mouth, and which in their form are much like bunches of currants, are always full of salivary secretion. There are nerves which are distributed from the brain to these glands, and when other nerves which belong to the senses of taste, of sight, or of feeling, are excited by the presence of food, *a stimulus* is imparted to the salivary glands, through the nerves that surround them, their cells collapse, and the juice which they contain is poured out through their stems, or ducts, into the mouth.

874. *How do we know that impressions imparted to one set of nerves, may be imparted to another set, so as to put any particular organ in action.*

Because very frequently *the mere sight* of rich fruit, or acid substances, *will cause the saliva to flow freely.* In this case it is evident that the salivary glands *could not see or know* that such substances were present. An impression must, therefore, be made upon the brain, *through the organ of vision,* and the desire to taste the substances being awakened, a nervous stimulus is *imparted to the glands of the mouth,* and they at once commence their action, *as if food were present.*

875. *Why does food descend into the stomach?*

Because, after the teeth, the tongue, and the muscles of the mouth generally, have rolled the food into a soft bolus, it is conveyed to the back of the mouth, where it is set upon the opening of the throat (*œsophagus*). It does not then descend through the throat by its own gravity, because the throat is generally in a compressed or collapsed state, like an empty tube; and we know that persons can eat or drink when with their heads downwards. The œsophagus is formed of a number of muscular threads, or rings, and *each little thread is like a hand ready to grasp at the morsel that is coming.* As soon as the bolus is presented at the top of the throat, these little muscular hands lay hold of it, and transmit it downward, passing it from one to another, until it is conveyed through the long passage, to the door of the stomach, which it enters.

"Remove far from me poverty and lies; give me neither poverty nor riches, feed me with food convenient for me."—PROVERBS XXX.

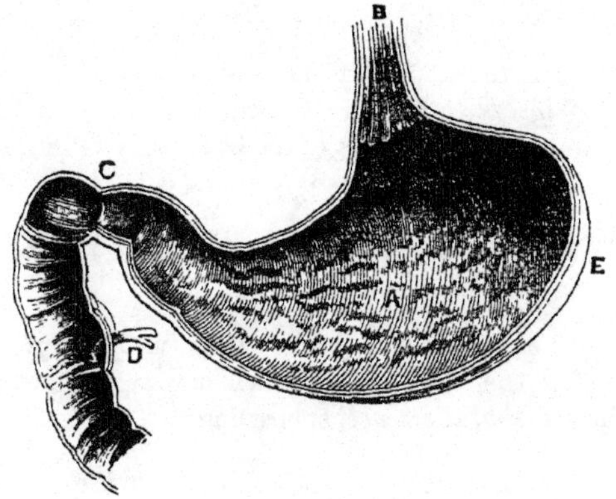

Fig. 49.—SECTION OF THE STOMACH, &c.

A. The inner coat of the *stomach.* (The stomach is here represented cut through its length, so that we can see its inside.)

B. The lower extremity of the throat, or *œsophagus,* through which food enters the stomach.

C. The passage out of the stomach, called the *pylorus,* where a muscular contraction prevents the escape of undigested food.

D. The *duodenum,* and the ducts through which the *bile* and *pancreatic* juices enter and mingle with our food.

876. *Why do we not feel the food being transmitted through the throat?*

Because the nerves of the body differ in their powers: some are nerves of *feeling,* some of *motion,* and others are nerves of the *senses.* The nerves of feeling are most abundantly distributed to those parts *where feeling is most useful and necessary to us.* But the faculty of feeling our food undergoing digestion would be no service to us whatever; therefore the nerves of *motion* are plentifully distributed to the throat and stomach, but very few of the nerves of *feeling*—just as many as will tell us when we eat anything *too hot,* or *too cold,* or that the stomach is *out of order.*

877. *Why do we feel uneasy after eating to excess?*

Because the stomach is *distended,* and presses upon the other organs by which it is surrounded.

"Who satisfieth thy mouth with good things; so that thy youth is renewed like the eagles."—PSALM CIII.

878. *Why do we feel drowsy after eating heartily?*

Because, while the stomach is in action, *a great proportion of the blood of the body is drawn towards it*, and as the blood is withdrawn from the other parts of the body, they fall into a state of languor.

879. *Why does blood flow more freely to the stomach during digestion?*

Because the energy of an organ is *increased by the flow of blood*, which supplies the *material* of which our organs are composed, and in which the *vital essence*, supporting life, resides.

880. *Why does excess in eating bring on indigestion?*

Because the power of the stomach to digest food is *governed by the amount of food required by the system.* It seems to be an instinct of the stomach to hold back food which is in excess, and by indications of pain and disturbance to warn its master that *excess has been committed.*

881. *Why is food digested in the stomach?*

Because it enters the stomach in the form of a paste, produced by the action of the mouth; and directly food enters, the *gastric juice*, which is formed by glands embedded in the coats of the stomach, trickles down its sides. This is a more *powerful solvent* than the salivary juice—it is like the same kind of fluid, only much stronger, and it soon turns the food from a rough and crude *paste* into a *greyish cream* (chyme). The heat of the stomach assists the operation, and the muscular threads of the coats move the cream along, in the same manner that the muscles of the œsophagus brought down the food.

The cream is passed towards the door which leads outward from the stomach (*pylorus*); but if, in the midst of the cream, there are any undissolved particles of food, it closes upon them, and they return again to the stomach to be further changed.

882. *Why does indigestion bring on bilious attacks?*

Because the *liver* secretes a fluid to assist in the digestion of food. The liver is a gland—a similar organ to the glands of the mouth—and it forms *bile* in the same manner that they form the salivary juice. Only the liver is a *much larger gland*, and a much greater quantity of blood passes through it. The liver pours its secretion into the biliary duct (Fig. 49) to mix with the grey cream as it passes onward, and to further dissolve it. But when the stomach is excited by food which it cannot dissolve, and when the owner of the stomach, disregarding its remonstrances, will persist in over-eating, or in eating things that disagree with the system, then *the liver and the stomach sympathise*, and the muscular threads, or hands, that prevail all through the alimentary organs, instead of moving *onward*, move backward, and *throw some bile into the stomach* to assist to dissolve and remove the excessive or improper food.

CHAPTER XLII.

883. *Why does some portion of the food we eat nourish the system, while other portions are useless?*

Because most food contains some particles that are indigestible, or that, if digested, are innutritious, and not necessary for the system. The *liver* is the organ by whose secretion the *useful is separated from the useless*; for when the bile enters through the duct (Fig. 49) and mixes with the grey cream coming from the stomach, it remains no longer a grey cream, but turns into a mass coloured by bile, having upon its surface *little globules of milk*, small, but very white. Those minute globules of milk (*chyle*) are the nutritious particles derived from the food; the other portion, coloured with bile, is the useless residue, or rather the *bulk from which the nutrition has been extracted.*

884. *Why does the milky, or nutritious matter, separate from the innutritious, upon admixture with bile?*

Because the bile contains an oily matter which *repels* the watery *milk of nutrition.*

"God hath made of one blood all nations of men for to dwell on all the face of the earth."—ACTS XVII.

The *pancreatic juice* also enters through the same duct with the bile. But its precise use is not understood. It is a fluid much like the salivary secretion of the glands of the mouth.

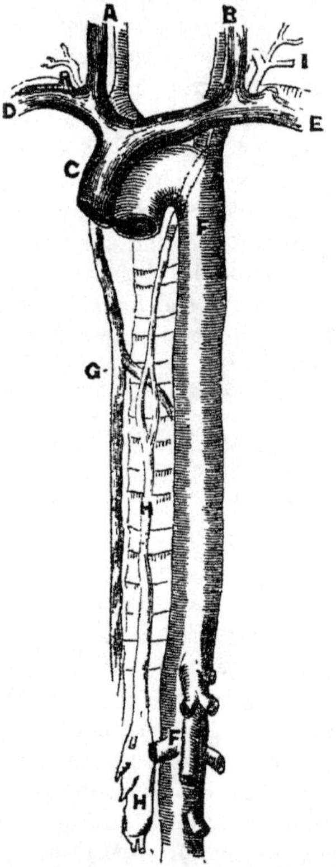

Fig. 50.—GREAT VESSELS OF THE CIRCULATION, AND THE DUCT WHICH CONVEYS NUTRITIVE MATTER TO THE BLOOD.

A B. *Jugular veins* which return blood from the head to the heart.

C. The *superior venæ cava*, or trunk vein, which pours the blood returned from the upper part of the system into the heart. There is a similar large vessel which meets this one and brings back blood from the lower part of the body, and they both pour the blood into the right side of the heart.

D E. The branches of the *venous system* which bring back the blood from the arms.

F F. The *great aorta*, the blood vessel which conveys arterial blood from the heart, and gives off branches that supply every part of the body.

G. Another large vein which returns the blood from the muscles of the chest, &c.

H H. The *thoracic duct*, which receives the newly dissolved food from the small absorbents, that collect it from the intestines. It conveys this nutrition (called chyle) upward along the back, until it reaches where the duct turns into the junction of two veins, and pours its contents into the veins bringing blood back to the heart. The nutrition, therefore, is at this moment mixed with the venous blood, and is sent to the lungs to be oxygenised.

885. *How is the nutrition taken away from the bilious residue?*

The muscular threads (or hands, as we figuratively call them) continue to push forward the digested matter through a long tube, called *the alimentary canal,* or bowels. This canal is some thirty feet in length, and is folded in various layers across the abdomen, and tied to the edge of a sort of apron, which is gathered up and fastened to the back-bone. All along this alimentary canal those muscular hands are pushing the digested mass along. But upon the coat or surface of the canal there are millions of little vessels called *lacteals,* which look out for the minute globules of milk as they pass, and *absorb* them, which means that they pick them up, and carry them away. There is an immense number of these little vessels, all busily at work picking up food for the system.

Then there is a large vessel, called the *thoracic duct,* which comes down and communicates with those little vessels (it is a sort of overseer, having a large number of workmen,) and collects the produce of their toil, and carries it upwards to the part where it passes *from the organs of digestion* into the *vessels of circulation.*

886. *What becomes of the nutrition, when it has entered the vessels of the circulation?*

It is sent through a large vein into *the heart,* entering that organ on the right side, from which the heart propels it into the lungs, mixed with *venous blood;* and the venous, or blue blood, is sent into the lungs, *taking with it the milk,* the formation of which we have traced.

887. *Why are the venous blood and the chyle sent to the lungs?*

Because the venous blood, in its circulation through the body, has parted with its *oxygen,* and taken up *carbon,* and it requires *to get rid of the carbon, and take up more oxygen.* The chyle, also, now combined with the blood, requires *oxygen,* and having obtained it, is converted into *bright red blood,* and the blue blood of the veins, having got rid of its carbon, which formed the carbonic acid of the breath, has again become *bright red blood.* We must therefore, in pursuing our description, cease to speak of blue, or *venous blood,* and of white milk, or *chyle,* for the two have now combined, and, with the oxygen of the air, have formed *arterial blood.*

888. *What becomes of the arterial blood thus formed?*

It is sent back from the lungs to the right side of the heart, from which it is sent into the *great trunk of the aorta*, and from thence it passes into smaller blood-vessels, until it finds its way to *every part of the system.*

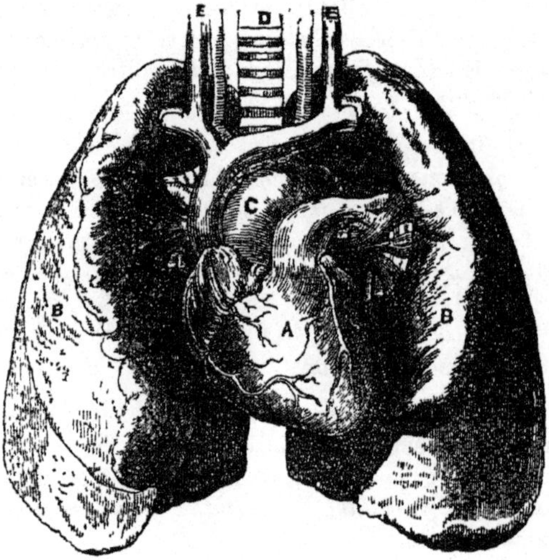

Fig. 51.—THE ORGANS OF RESPIRATION.

A. The *heart.*

B B. The *lungs.*

C. The *aorta*, and on either side of the aorta the vessels which convey the venous blood to the lungs to be *oxygenised*, and the corresponding vessels which return it to the heart, after it has undergone that operation. (For *aorta* see Fig. 50.)

D. The *trachea*, or large air passage, through which the air passes into the spongy texture of the lungs, when we breathe.

E E. *Arteries* and *veins*, being the trunks of the vessels that supply the head, &c.

889. *Why does the chest expand when we breathe?*

Because the lungs consist of *millions of hollow tubes*, and *cells*, which, having been emptied by throwing off *carbonic acid gas* and *nitrogen*, become compressed, and the atmospheric air flowing into these millions of spaces, and filling the lungs, just as water fills and swells a sponge, causes them to expand, and occupy greater room.

890. *How does the blood communicate with the air in the lungs?*

Through the *sides of very minute vessels*, of which, perhaps, a *fine hair* gives us the best conception. But these vessels are *twisted and wound round each other* in such a curious manner, that they form *millions of cells*, and by being twisted and wound, a much *greater surface of air and blood* are brought to act upon each other, than could otherwise be accomplished.

891. *Why does the blood which is thus formed, impart vitality to the parts to which it is sent?*

Because the blood is itself *vitalised*—is, in fact, *alive*, and capable of diffusing life and vitality to the organisation of which it forms a part.

This is a very wonderful fact, but no less true than wonderful, that dead matter which, but a little while ago, was being ground by the teeth, softened by the saliva, and solved by the gastric juice and bile, has now acquired *life*. Nobody can tell the precise stage or moment when it began to live. But somewhere between the stomach and the lungs, melted by the gastric juice, softened by the secretion of the pancreas, separated by the bile of the liver, macerated by the muscular fibres of the bowels, taken up by the absorbents, warmed by the heat of the body, and ærated in the lungs, it has by one, or by all of these processes combined, been changed from the dead to the living state, and now forms part of the *vital fluid of the system*.

CHAPTER XLIII.

892. *Why do we know that the blood has become endowed with vital powers?*

Because, in the course of its formation, it has not only undergone change of condition and colour; but, if examined now by the microscope, it will be found to consist of millions of minute cells, or discs, which float in a watery fluid. The paste produced by mastication consisted of a crude admixture of the atoms of food; the cream (*chyme*) formed from this in the stomach, presents to the microscope a heterogeneous mass of matter, exhibiting no appearance whatever of a new organic arrangement; the milk (*chyle*) which is formed in the intestines is found to contain a great number of very small molecules, which probably consist of some fatty matter; as the chyle progresses towards the *thoracic duct* (Fig. 50), it appears to contain more of these, and slight indications present themselves of the approach towards a new organic condition.

But wherever *vitalisation begins*, no human power can say with confidence. Yet there can be no doubt that the blood is both *organised* and *vitalised*, and that it consists of corpuscles, or little cells, enclosing matters essential to life.

893. *Why does the blood circulate?*

Because all the bones, muscles, blood-vessels, nerves, glands, cartilages, &c., of which the body is composed, are constantly undergoing a *change of substance*. It is a condition of their life, health, and strength, that they shall be "*renewed*," and the blood is the great source of the *materials* by which the living temple is kept in repair.

894. *How is the body renewed by the blood?*

Every drop of blood is made up of a large number of corpuscles, each of which contains some of the elements essential to the wants of the system.

Let us, to simplify the subject, consider the blood vessels of the body to be so many *canals*, on the banks of which a number of inhabitants live, and require constant sustenance. The corpuscles of the blood are the *boats* which are laden with that sustenance, and when the heart beats, it is a signal for them to start on their journey. Away they go through the arch of the great *aorta*, and some of the earliest branches which it sends off convey blood to the arms. We will now for a moment dismiss the word *artery*, and keep up the figure of a system of canals, with a number of towns upon their banks.

Well, away go a fleet of boats through the *aorta* canal, until they reach a point which approaches Shoulder-town; some of the

"Though hand join in hand, the wicked shall not be unpunished; but the seed of the righteous shall be de ivered."—PROVERBS XXI.

boats pass into the *axillary* canal and Shoulder-town is supplied; the other boats proceed along the *humeral* canal until they approach Elbow-town, when another division of the boats pass into other branch canals and supply the wants of the neighbourhood; the others have passed into the *ulnar* canals and the *radial* canals until they have approached Wrist-town and Hand-town, which are respectively supplied; and then the two canals have formed a junction across the palm and supplied Palm-town, where they have given off branches and boats to supply the four Finger-towns, and Thumb-town.

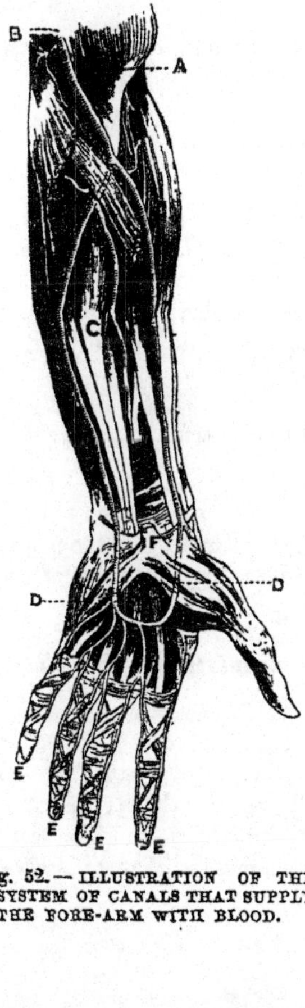

Fig. 52.—ILLUSTRATION OF THE SYSTEM OF CANALS THAT SUPPLY THE FORE-ARM WITH BLOOD.

Between A and B the *brachial canal*, which gives off branches to supply Elbow-town, &c., and then divides into two main courses, diverging to the opposite sides of the arm, and sending a smaller canal down the centre.

D D. The point where the *ulnar canal* and the *radial canal*, after having passed and supplied Wrist-town, form a junction, running through Palm-town, and in their course giving off branches to supply the four Finger-towns and Thumb-town.

For further explanations of the engraving, see 57.

895. *How does the blood return to the lungs, after it has reached the extremities?*

The *veins* constitute a system of vessels corresponding to the arteries. We may say that the arteries form *the down canal*, and the veins *the up canal*. The arteries, commencing in the great trunk of the *aorta*, branch off into large and then into smaller tubes, until they form *capillary* or hair-like vessels, penetrating everywhere.

The capillary extremities of the arteries, unite with the capillary extremities of the veins, and the blood passes from the one set of vessels into the other. As the *arteries become smaller* from the point where they receive the blood, so the *veins grow larger*; the venous capillaries, pour their contents into small vessels, and these again into larger ones, until the great *venous* trunks are reached, and the blood is passed again into the heart as at first described. (Fig. 50.)

896. *Why do we see blue marks upon our arms and hands?*

Because large veins lie underneath the skin, through which the blood of the fingers and hand is *conveyed back to the heart.*

897. *Why are the veins more perceptible than the arteries?*

Because the arteries are buried *deeper in the flesh, for protection.* It would be *more dangerous to life to sever by accident an artery than a vein.* A person might bleed longer from a vein than from an artery, without endangering life; because the arteries supply the *life sustaining blood.* The Almighty, therefore, has buried the arteries for safety.

898. *Why when we prick the flesh with a needle does it bleed?*

Because the capillary arteries and veins are so fine, and are so thickly distributed all over the body, that not even the point of a needle can enter the flesh without penetrating the coats of several of these small vessels.

899. *What occurrs during the circulation of the blood?*

Not only do the various parts to which the boats are sent take from them whatever they require, but *the boats collect all those matters for which those parts have no further use.* The bones, the nerves, the muscles, &c., all renew themselves as the boats pass along; and all give something to the boats to bring back. One of the chief exchanges is that of *oxygen* for *carbon*, by which a gentle *heat* is diffused throughout the system. It is for this purpose that *fresh air* is so *constantly necessary*.

But other exchanges take place. The blood, in addition to oxygen and carbon, contains *hydrogen* and *nitrogen*. But it contains its four elements in *various forms of combination*, producing the following *materials* for the use of the body: of 1,000 parts of blood, *about* 779 are *water*; 141 are *red globules*; 69 are *albumen*; 3 are *fibrin*; 2 are *fatty matter*; 6 are various *salts*.

Albumen and fibrin are a kind of flesh imperfectly formed, and probably are chiefly used in repairing the muscles. The red corpuscles contain the oxygen which goes to combine with the superabundant carbon, and develope heat; the fatty matters probably repair the fatty tissues, and glands that are of a fatty nature; and the various salts contribute to the bones, and to the chemical properties of those secretions which are formed by the glands, &c., while the great proportion of water is employed in cleansing, softening, and cooling the whole, or the living edifice, and it is the medium through which all the nutrition of the body is distributed.

900. *Why do we feel the pulse beat?*

Because every time that the heart contracts it send a fresh supply of blood to the blood-vessels, and the motion thus imparted creates *a general pulsation throughout the system*: but it is more distinctly perceived at the pulse, because there a rather *large artery lies near to the surface*.

901. *What becomes of the matter collected by the blood in the course of its circulation?*

We have already explained that carbon is thrown off from the lungs in the form of carbonic acid gas. But there are many other matters to be separated from the venous blood, and its putrification is assisted by the action of the liver, which is supplied with a large vein, called the *portal vein*, which conveys into the substance of the liver, a large proportion of the venous blood, from which that organ draws off those matters which form the bile, and other matters which are transmitted with the bile to the bowels. The *liver* and

"Thy hands have made me and fashioned me: give me understanding, that I may learn thy commandments."—PSALM CXIX.

the *lungs*, therefore, are the great purifiers of the venous blood. But there are also smaller organs that assist in the same work.

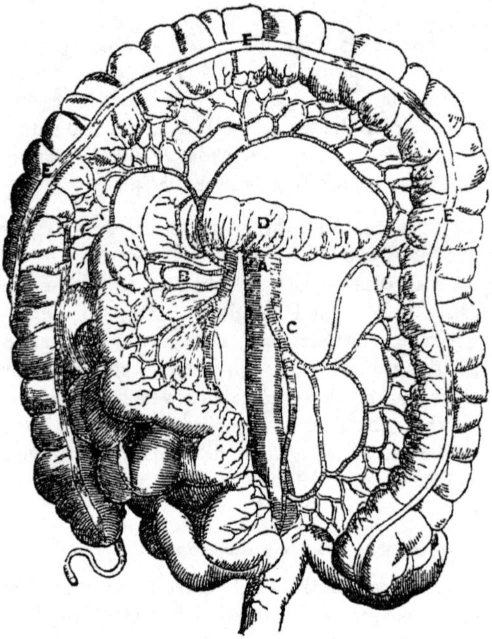

Fig. 53.—SHOWING THE DISTRIBUTION OF BLOOD THROUGH BRANCHES OF THE AORTA.

A. The *aorta*.

B. Branches given off for the *aorta* to supply one portion of the intestines.

C. Branches given off by the aorta to supply other portions of the intestines. A complete communication may be traced between these vessels from the origin of one to that of the other.

D. The *pancreas*, or sweetbread, a large gland that forms the pancreatic juice. which it pours in through the duct. *See* Fig. 50.

E E E. The *large intestines*, forming the termination of the alimentary canal.

CHAPTER XLIV.

902. *Why when we cut our flesh does it heal?*

"And God said, Let us make man in our own image, after our likeness; and let them have dominion over the fish of the sea, and over the fowl of the air, and over the cattle, and over all the earth, and over every creeping thing."—GEN. I.

Because the blood coagulates over the cut, and throws out a kind of *lymph*, which forms an incipient flesh, and excludes the air while the blood-vessels are engaged in *repairing the part.*

903. *Why, since all the substance of the body undergoes change, do we preserve the same features throughout our lives?*

Because our substance changes in the *minutest atoms;* and each separate atom has a life of itself, the maintenance of which preserves the *unity and permanence of the whole.*

904. *Why do moles upon the skin continue permanent, while bruises and wounds disappear?*

Because moles are themselves *organised formations,* and repair themselves just as any other part of the body does. But bruises and wounds are the result of *accidental disturbances,* which in course of time become removed.

905. *Why do the marks of deep cuts sometimes remain?*

If the cut is so deep and serious as to destroy the *system of vessels* which supply and repair the part, then it is evident that they cannot work so perfectly as when in their sound condition. Their functions are, therefore, interfered with, and instead of having flesh uniform with the other parts of the system, there results a *scar*, or a wound *imperfectly repaired.*

906. *Why when we hold our hands against a candle-light do we perceive a beautiful crimson colour?*

Because the fluids and vessels of the body are in some degree transparent, and the thin textures of the sides of the fingers allows the light to pass, and shows the beautiful crimson colour of the blood.

If the web of a frog's foot be brought in the field of a good microscope, and set against a strong light, the blood may be seen in circulation, with the most wonderful effect. Each vessel, and every globule of blood, can be seen most distinctly, and the junction of the arteries and veins can be clearly traced. The little boats of nutrition may be seen chasing each other in rapid succession, and when the animal exerts itself to escape, the flow of the blood increases; and not unfrequently, under these circumstances of agitation, have we seen two or three blood discs struggling together to enter a vessel that was too small for them. Again and again they have endeavoured to find a passage, until one of them happening to slip forward got away, followed by the others!

"Know ye that the Lord he is God: it is he that hath made us, and not we ourselves: we are his people, and the sheep of his pasture."—PSALM C.

907. *Why does the flesh underneath the nails look red?*

Because the transparent texture of the nails enables us to see the colour of the *vascular structure* that lies underneath the skin.

Vascular.—Full of vessels. In this instance, full of capillary blood-vessels.

908. *Why have we nails at our fingers' ends?*

Because they give *firmness to the touch*, and enable us to apply the extremities of the fingers to many useful purposes for which they would otherwise be unfitted. They enable us to *press the tips of the fingers*, where the highest degree of sensitiveness prevails, so as to bring *the largest amount of nervous perception* into the sense of touch.

909. *Why do white spots occur upon the nails?*

Because the vascular surface underneath is attached to the horny texture of the nail; but by knocks and other causes, the nail sometimes *separates in small patches from the membrane* below, and becomes *dry and opaque.*

910. *Why is there a circular line of whitish colour at the root of the nail?*

Because there the nail is *newly formed* by the vascular substance out of which it grows, and has not yet assumed its proper horny and transparent nature.

911. *Why is the eye-ball white?*

Because the blood-vessels that supply its surface are so very fine that they do not admit the *red corpuscles* of the blood.

912. *Why does the eye-ball sometimes become blood-shot?*

Because, under exciting causes of inflammation, the *blood-vessels become distended,* and the red corpuscles enter, producing a network of red blood-vessels across the white surface of the eye.

913. *Why are the lips red?*

Because the lips are formed of the *mucous membrane* that lines the body internally, and covers the surface of most of the internal parts. This membrane contains a great number of minute red vessels, which give softness and moisture to the surface. A very beautiful illustration of the softness, moisture, and delicate colour of the mucous membrane is afforded by turning up and examining the under surface of the upper eye-lid.

914. *Why do delicate persons look pale and languid?*

Because, generally from the want of exercise and fresh air, their blood is deficient of the healthy proportion of *red corpuscles.*

915. *Why does exercise and fresh air impart to healthy persons a red and fresh appearance?*

Because the redness of the blood is due to the *amount of oxygen* which it contains, and air and exercise *oxygenise* the blood, and diffuse it throughout the system.

916. *How is the blood propelled through the arteries?*

By the very powerful contraction (and alternate dilation) of the thick *muscles of the heart,* assisted also by the *muscular cords of the blood-vessels* themselves, and in many instances by the *compression of the muscles* in which the arteries lie embedded.

917. *Why are the capillary arteries capable of receiving the great quantity of blood sent out through the larger vessels?*

Because the capillary vessels are *so numerous,* that though they are infinitely smaller, they are capable of receiving in their minute tubes *the whole of the quantity of blood* transmitted to them through the larger vessels.

918. *Why, when we sit with our legs crossed, do we see the foot that is raised move at regular intervals?*

Because the pressure upon the muscles of the leg retards the progress of the blood until it forces *itself through the compressed vessels,* and thereby imparts a pulsation which moves the leg and foot.

919. *Why are capillary blood-vessels found in every part of the system?*

Because it is *through these small vessels alone* that the substances of the body are renewed and changed. Even the larger blood-

"All my bones shall say, Lord, who is like unto thee, which deliverest the poor from him that is too strong for him, yea, the poor and the needy from him that spoileth him?"—PSALM XXXV.

vessels *do not sustain themselves upon the blood which they contain*, but receive into their coats numerous capillary vessels by which they are nourished.

920. *How much blood does the human body contain?*

From *twenty-five* to *thirty-five* pounds. (*See* 623.)

921. *How does the blood ascend in the veins, in opposition to gravitation?*

In addition to the muscular coats of the veins, and the influence of muscular action upon them, there are in the veins numerous semi-circular valves, which are not found in the arteries. These valves extend from the sides of the veins in such a manner that they allow the free passage of the blood upwards, but a backward motion of the blood would expand the cup-like valves and stop the passage; so that the blood can only move in one direction, and that *towards the heart.*

922. *How frequently does the total amount of blood circulate through the system?*

The blood circulates once through the body in about *two minutes.* If, therefore, we estimate the amount of blood at twenty-four pounds, it follows that no less than *twelve pounds of blood pass through the heart every minute*; and it is estimated that if the blood moved with equal force in a straight line it would pass through *one hundred and fifty feet in a minute.*

CHAPTER XLV.

923. *How many bones are there in the human body?*

There are *two hundred and forty-six*, and they are apportioned to the various parts of the body in the following numbers:—

Head	8
Ears	6
Face	14
Teeth	32
Back-bone and its base	26
Chest, &c.	26

Arms and Hands	64
Legs and Feet	62
Small moveable bones	8

924. *Of what substances are the bones composed?*

One hundred parts of bone consist of

Cartilage	32·17	parts
Blood-vessels	1·13	,,
Carbonate of lime	11·30	,,
Phosphate of lime	51·04	,,
Fluate of lime	2·00	,,
Phosphate of Magnesia	1·16	,,
Soda, chloride of sodium	1·20	,,
	100·00	

925. *What are the uses of the bones?*

They *protect* soft and delicate organs; they form a framework to which the organs are attached, and by which they are *kept in their places;* and they supply a *mechanism,* by which the *motions of the body are produced,* in combination with the muscles.

926. *Why is the brain placed within the skull?*

Because that delicate and vital organ, being the *centre and the root of the nervous system*, requires a position of the *greatest safety.*

927. *Why are the bones that constitute the vertebræ (back-bone) hollowed out, so as to form a continuous groove?*

Because through that groove the *spinal cord* passes out from the brain. Being in the centre of that column of bones, the spinal cord receives from them a similar protection *to that which the brain obtains from the skull.*

928. *Why is the head set upon the neck?*

Because in that position it obtains the *freest motion,* can turn in *any direction,* and is placed relatively to the other parts of the body, in that situation where it acquires *the greatest possible advantage.*

929. *Why are the eyes placed in the sockets of the skull?*

Because the bones of the skull *afford protection* to the delicate and complicated structure of the eyes, and supply points of attachment, and grooves, by which the muscles are enabled to *turn the eyes freely*, and thereby *extend the field of vision.*

930. *Why are the bones of the skull arched?*

Because in that form they acquire *greater strength*, and hence the utmost degree of safety is combined with extreme *lightness of material.*

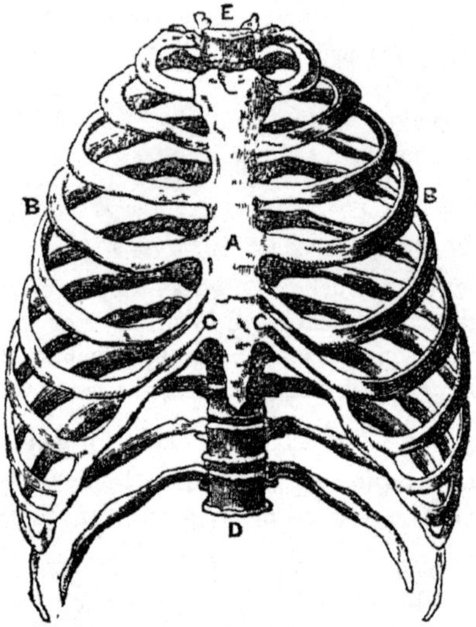

Fig. 54.—VIEW OF THE BONES OF THE THORAX, OR CHEST, SHOWING THE PROTECTION AFFORDED TO THE ORGANS OF CIRCULATION AND RESPIRATION.

A. The sternum, or breast-bone.

B B. The *ribs*, which rise a little from behind, and fall as they come forward, by which they acquire a greater flexibility.

C C. The *cartilaginous points* of the short ribs, by which their expansive and compressive powers are much increased.

D E. Part of the vertebral column, or back-bone.

931. *Why are the bones of the skull divided by sutures (seams), with points which fit into each other like small teeth?*

Because, by that arrangement, *concussions of the skull*, which might be fatal to the brain, are *deadened*, and injuries from accident *greatly modified.*

"And I will lay the sinews upon you, and will bring up flesh upon you, and cover you with skin, and put breath in you, and ye shall live; and ye shall know that I am the Lord."—EZEKIEL XXXVII.

932. *Why are the heart, lungs, &c., placed within the chest?*

Because the functions of those organs require *considerable space*, while their importance in the system of life, renders it essential that they should be *securely protected* from the probabilities of accident.

933. *Why are the heart and lungs enclosed for protection in a series of ribs, and not in a close case, like the brain?*

Because, by the inflation and contraction of the lungs, their *capacity is constantly changing.* When man takes a moderate inspiration, he inhales about *thirty cubic inches of air*, and the lungs increase in size *one-eighteenth of their whole capacity.* Consequently, were they enclosed in a frame of *fixed dimensions*, it must needs be, to that extent at least, larger than is necessary, when the frame is made to dilate and contract with the capacity of the lungs

So perfect is the Almighty contrivance, that not only are the ribs made to *protect* the lungs, but, by their elasticity, and the contractions and dilations of the muscles which lie between them, they *assist the lungs in their labours*, and work with them in perfect harmony.

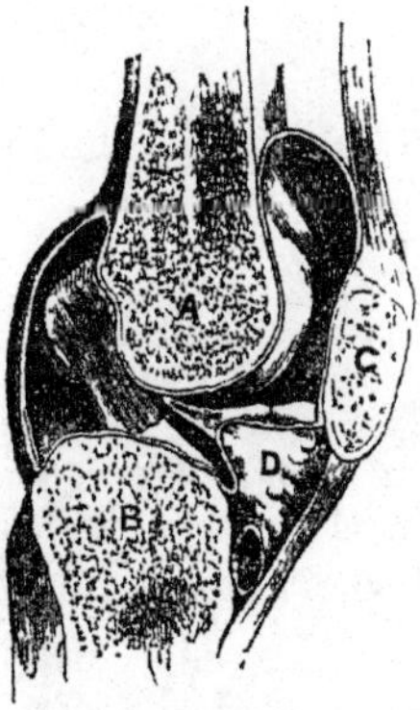

Fig. 55.—SECTION OF THE KNEE JOINT, SHOWING THE CELLULAR STRUCTURE OF BONE, BY WHICH LIGHTNESS AND STRENGTH ARE OBTAINED.

934. *Why are the bones of the arms, legs, &c., made hollow?*

Because *lightness* is thereby combined with *strength.* There is a provision by which, in the extremities of bones, where an enlarged surface is required, *lightness* is still combined with the necessary degree of strength.

The bones are made up of a *cellular formation;* and this generally occurs in parts which are much called into action, in the various movements of the body.

A. Lower part of the bone of the thigh.

B. Head of the bone of the leg.

C. The knee cap, showing its relation to the other bones, and the manner in which it is enclosed by the tendons seen at Fig. 58.

D. A pad of fat, lessening the friction of the bones, and modifying the shocks produced by jumping, &c.

"Again he said unto me, Prophesy upon these bones, and say unto them, O ye dry bones, hear the word of the Lord."—EZEKIEL XXXVII.

935. *Why are the bones of the arms and legs formed in long shafts?*

Because a considerable *leverage* is gained, by which the advantages of *quickness of motion*, and *increase of mechanical power*, are secured.

936. *Why are the bones of the hands and feet numerous and small?*

Because the motions of the hands and feet are very *varied and complicated*. There are no less than *twenty-eight bones* in one hand and wrist; and about *as many* in a foot and ankle. To these are fastened a great number of *ligaments and muscles*, by which their varied compound movements are controlled. But for the complexity of the mechanism of our hands and feet, our motions would be extremely awkward, and many of the valuable mechanical inventions which now benefit mankind, could never have been introduced. The bones of the hands and feet are in number equal to *one-half of the whole of the bones of the body*.

CHAPTER XLVI.

937. *What are ligaments?*

Ligaments consist of bands and cords of *a tough, fibrous, and smooth substance*, by which the bones are bound together and held in their places, allowing them freedom to move, and supplying smooth surfaces over which they glide.

938. *Why are the joints bound with ligaments?*

Because the bones would otherwise be constantly liable to *slip from their places*.

"That which is born of the flesh is flesh; and that which is born of the Spirit is spirit."—JOHN III.

939. *What are tendons?*

Tendons are *long cords*, of a substance similar in its nature to *cartilage*, by which *the muscles are attached to the bones.*

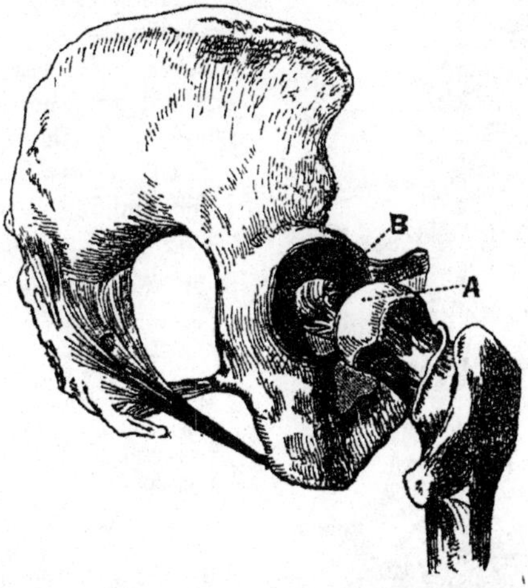

Fig. 56.—SHOWING A BALL AND SOCKET JOINT, AND THE MANNER IN WHICH LIGAMENTS ARE EMPLOYED TO HOLD BONES IN THEIR POSITIONS.

A. The *ball*, or *head* of the thigh bone.
B. The *socket*, showing the ligament in the socket, which holds the head of the bone in its place, but allows it free motion.
C. *Ligaments* tied from bone to bone, giving firmness to the parts.

940. *Why are tendons used to attach the muscles to the bones?*

Because, by this arrangement, the large muscles by which the extremities are moved, *may be placed at some distance* from the bones upon which they act, and thus the extremities, instead of being large and clumsy, are *small* and *neat.*

941. *How many muscles are there in the human body?*

There are about *four hundred and forty-six muscles* that have been dissected and described, and the actions of which are perfectly understood. But there is probably a much larger number of muscles, and of compound actions of muscles, than the skill of man has been able to recognise.

"All flesh is not the same flesh: but there is one kind of flesh of men, another flesh of beasts, another of fishes, and another of birds."—CORINTHIANS XVI.

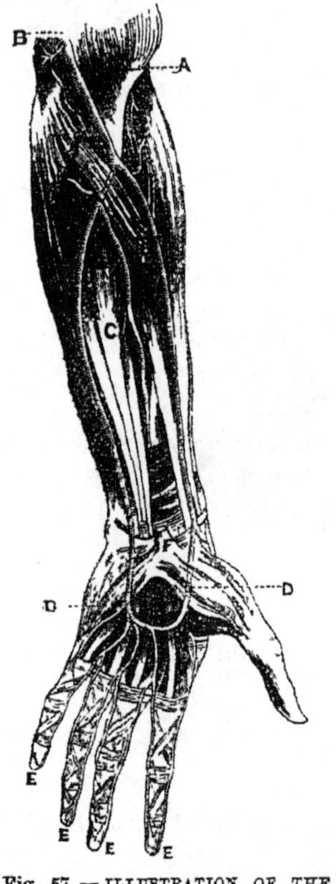

Fig. 57.—ILLUSTRATION OF THE RELATION OF MUSCLES, TENDONS, AND BONES.

942. *What is the constitution of a muscle?*

Every muscle is made up of a number of *parallel fleshy fibres*, or threads, which are bound together by a smooth and soft tissue, forming a sheath or case to the muscle, and enabling it to *glide freely* over the surfaces upon which it moves.

A. Lower extremity of the muscle which draws the fore-arm towards the upper-arm, bends the elbow, raises the hand to the head, and is powerfully exerted in pulling, lifting, &c.

C. A muscle which gives off four long *tendons*, which pass under the *ligaments* of the wrist, one to each finger, and by which the fingers are bent upon the palm of the hand, as in grasping, &c.

F. *Tendon* of a muscle which draws the little finger and the thumb towards each other.

The *ligaments* may be seen enfold- the finger-joints, and also crossing the wrist, underneath the *tendons*.

The muscles are compressed into *tendinous cords* at their ends, by which they are *united to the bones*.

They are arranged *in pairs*, having reciprocal actions—each muscle having *a companion muscle* by which the part which it moves is restored to its original position, when the influence of the first muscle is withdrawn, and the stimulus given to bring back the part.

943. *Why can we raise our fingers?*

Because muscles which lie *on the fore-arm*, and have their

tendons fastened at the ends of the fingers, *contract*, and by becoming shorter, *draw the fingers upward*, and towards the arm.

944. *Why can we throw back the fingers after they have been raised?*

Because the muscles at the back of the arm, *whose tendons are attached to the back of the fingers*, contract and restore them to their former position.

945. *What degree of strength do the muscles possess?*

The degree of strength of a muscle depends upon the *healthy condition* of the muscle, the *amount of stimulus* which it receives at the time of exertion, and the manner in which *its powers are applied*.

The great muscle of the calf of the leg has been found, when removed from a dead body, to be capable of sustaining a weight equal to *seven times the weight of the entire body*

But the contractile power of the living muscles is very great: the thigh bone has frequently been broken by muscular contractions in fits of epilepsy. And in cases where there has been a dislocation of the thigh, the head of the thigh-bone being thrown out of its socket, (Fig. 56) it has been found necessary to employ strong ropes, attached to a wheel turned by several hands. in order to *overcome the contraction of the excited muscles, and to enable the operator to restore the bone to its place.*

946. *What is the stimulus which sets the muscles in action?*

The muscles are excited to action by *the nerves*, which they receive from the *spinal cord.*

947. *Why does it require the influence of the will to set the arms in motion?*

Because the muscles which form their mechanism are *voluntary* muscles—that is, they are subject to the *will of man*, and influ-

"And he took him by the right hand, and lifted him up; and immediately his feet and ancle bones received strength."—ACTS III.

Fig. 58.—MUSCLES AND VESSELS OF THE LEG AND FOOT.

enced by impulses directed to them through the nervous system *by the mind*, which is the governing power.

948. *Why does the heart beat without any effort of the will?*

Because the muscles of the heart are *involutary muscles*—that is, they are *independent of the will*, and receive *a continuous nervous stimulus* which is not *under the controul of the mind.*

A. A large *ligament*, which covers the knee pan, or moveable bone of the knee, by which the ends of the bones of the thigh and leg are kept from slipping over each other.

B. A muscle which passes underneath the cartilages of the ankle, and gives off four *tendons*, which are distributed to the toes, and by which they are extended in elongating the foot, walking, &c.

C. Part of the muscle which forms the fleshy bulb of the calf of the leg, and which terminates in the large *tendon* attached to the heel, called the *tendon of Achilles*.

D. One of the ligaments which bind the tendons and the bones of the ankle.

E. *Arteries* proceeding from the large vessel descending the leg, by which the toes are supplied.

949. *Why are the muscles of the arms, &c., made subject to the controul of the will?*

Because, as they supply the mechanism through which we adapt ourselves to our varying wants and circumstances, it was necessary that they should be placed under the controul of the mental power, and be moved *only in accordance with man's necessities.*

950. *Why are the motions of the heart, &c., made independent of the will?*

Because, as the necessity for the heart's motion is *fixed and unalterable,* the constant motion of the heart could be best secured by giving it a *fixed nervous influence,* by which it might be unfailingly prompted to fulfil its functions.

If the movements of man's heart were *subject to his will,* he would be constantly required to regard the operations of that organ; and so large an amount of mental care and physical exertion would have to be employed in that direction, that man's sole work would be to keep himself alive. Hence we see the goodness of the Creator in *giving* life to man, and in *keeping the vital impulses under his divine care.*

CHAPTER XLVII.

951. *What are nerves?*

The nerves are branches of the *brain* and the *spinal cord;* they are distributed in great numbers to all the active and sensitive parts of the body.

952. *What is the spinal cord?*

The spinal cord is a long and large cord of nervous matter, which extends from the brain through a continuous tube formed by corresponding hollows in the bones of the back. It serves as a nervous trunk *for the distribution of nerves,* just as the aorta distributes branches of blood-vessels.

953. *Why is the spinal cord placed in the grooves formed by the back-bone?*

Being a very vital part of the system, and from the delicacy of its structure liable to injuries, it is set in the back-bone for *protection;* and so great is its security that it is only by force of an unusual kind that it can be injured.

"A sound heart is the life of the flesh: but envy is the rottenness of the bones."—PROVERBS XIV.

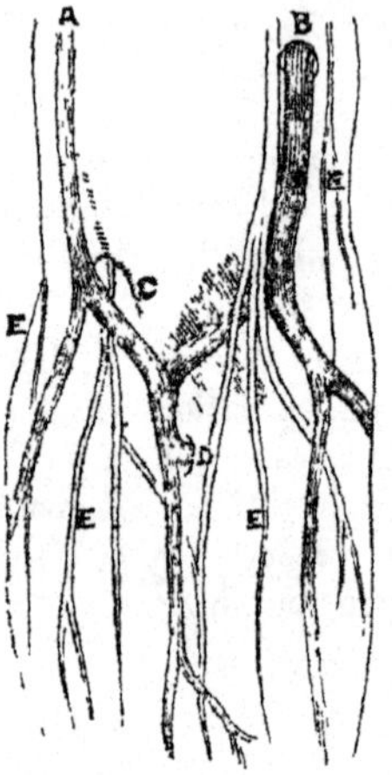

Fig. 59.—SHOWING THE DISTRIBUTION OF NERVES AND VEINS, AND ILLUSTRATING THE MANNER IN WHICH THEY PASS THROUGH THE FLESH TO REACH THE PARTS TO WHICH THEIR FUNCTIONS BELONG.

954. *How can branches proceed from it, if it is so securely encased in bone?*

Because in the bones, on each side of of the spinal cord, there are *smaller grooves* for the transmission of the nervous branches.

955. *Of what does the nervous system consist?*

Of the *brain*, the *spinal cord*, and the branches which are called *nerves*.

A B. *Veins* of the fore-arm.

B. Canal formed in the muscle, through which a *trunk-vein* emerges.

C. Canal formed in the muscle, through which a large *nerve* emerges.

D. Canal through which a *vein* enters to communicate with the deep muscles of the arm.

956. *What is the constitution of a nerve?*

It consists of a thin membrane, or sheath, surrounding *a greyish oily matter*, which forms the nervous marrow. In the centre of this marrow is usually found *a small fibre*, which is supposed to be the essential part of the nerve; and most nerves consist of a number of these sheaths enclosing fibres running in parallel directions.

957. *What is the nervous fluid?*

The term *nervous fluid* is used to express our ideas of the mode by which the brain and spinal cord influence the remote parts: just as we say the *electric fluid*, without knowing that such a fluid exists. It is the most convenient form of expression.

958. *How many classes of nerves are there?*

There are:—

1. The nerves of *motion*.
2. The nerves of *sensation*.

"Having many things to write unto you, I would not write with paper and ink; but I trust to come unto you, and speak face to face, that our joy may be full."—II JOHN.

3. The nerves of *special sense.*
4. The nerves of *sympathy.*

959. *What are the nerves of motion?*

The *nerves of motion* are those which, in obedience to the will, *stimulate the muscles to act,* and apportion the amount of stimulation they convey to the degree of exertion required.

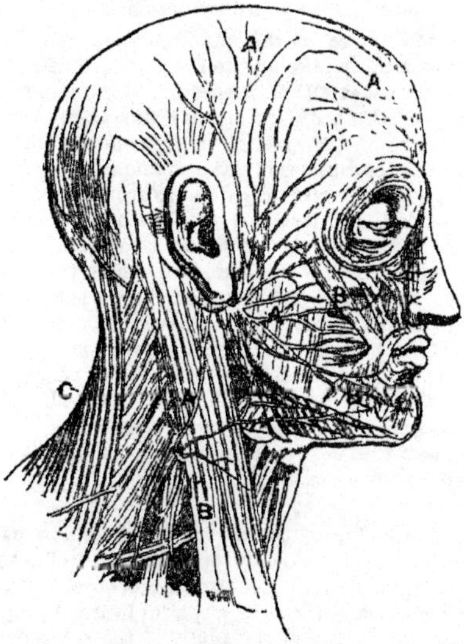

Fig. 60.—MUSCLES OF THE HEAD AND FACE, WITH NERVES DISTRIBUTED THERETO.

A A A. The *facial nerve* emerging from underneath the ear, and distributing branches to the cheeks, temple, forehead, &c. This nerve excites the muscles of the face, and is chiefly instrumental in producing the expressions of the countenance under the changing emotions of the mind.

B B B. *Muscles* by which various motions are imparted to the head, face, mouth, &c., under the *stimulus of the nerves.*

960. *What are the nerves of sensation?*

The *nerves of sensation* are those which *impart a consciousness to the brain* that its commands to the nerves of motion have been obeyed, and how far they have been fulfilled.

"Oh that men would praise the Lord for his goodness, and for his wonderful works to the children of men."—PSALM CVII.

Let us perform a simple experiment, which will more clearly illustrate the phenomena of *motion* and of *sensation*, which we are now describing, than a great deal of writing upon the subject. You hold in your hand this book: close it, and set it upon the table; lay your hands passively upon your lap, and then *will* your hand, to take up the book, which is the same as to say, *command* your hand to take up the book. What occurs? The hand, immediately obeying your desire, stretches forward to the book, and takes hold of it. How do you know that you have hold of it? You *see* that you have: but were your eyes closed, you would be equally aware that the hand had reached the book, and fulfilled your wishes. It is by the nerves of *sensation* that you are made aware that the hand has fulfilled your instructions.

Consider what took place in the simple action. In the first instance, a desire arose in your mind to take up the book. The *brain* is the organ of the mind; and having branches either proceeding from itself, or from the spinal cord, to every part of the body—branches that traverse like telegraphic wires throughout every part of the system,—it transmitted instructions along the nerves that proceed to the muscles of the arm and hand, directing them to take up the book. This was done instantly; and as soon as it was done you became conscious that your will had been obeyed—because *the nerves sent back a sensation to the brain acquainting it that the book had been taken up*, and that at the moment of the dispatch it was in the firm hold of the hand.

In all the varied motions of the body this double action of the nerves takes place. It is obvious that without an *outward* impulse from the brain, upon which the desire of the mind first made an impression, no motion of the muscles of the arm and the hand could have taken place; and it is also obvious that without an *inward* impulse from the nerves to the brain you would not have known that the muscles had fulfilled your instructions. The hand might have dropped by the side of the book, or have gone too far, or not far enough, and you would not have been waare of the result, but for an inward communication through the nerves.

We are not now speaking of the nerves which endow us with the sense of *feeling;* because they are regarded as separate and distinct from those nerves that produce in us consciousness of muscular response. When we walk, rise, or sit, we are made conscious, without any special feeling being exerted, that the muscles have placed the limb, or the body, in the desired position, that it is set down safely and firmly, and that we may repose upon it securely without further attention. We refer the impressions made by the book upon the nerves of the hand, and which enable us to tell whether it feels hot or cold, whether its surface is rough or smooth, and so on, to the special sense of *feeling*. The consciousness of muscular action is a separate and distinct function; and it is generally believed that the same nerves that convey the command of the will outward, bringing back the intimation that the will has been obeyed, but that *different fibres* of the nerves convey the *outward* and the *inward* impulses. A single nerve may therefore be likened to a *double wire* connected with the electric telegraph: one transmitting despatches in one direction, and the other in the opposite direction.

961. *What are the nerves of special sense?*

The nerves of special sense are those through which we *hear*, *see*, *feel*, *smell*, and *taste*.

"For the Lord seeth not as man seeth; for man looketh on the outward appearance, but the Lord looketh on the heart."—SAMUEL XVI.

962. *What are the nerves of sympathy?*

The nerves of sympathy, or the system of *sympathetic nerves*, are those which are distributed to the *internal organs*, and which are independent of the will. They regulate the motions of the heart, the lungs, the stomach, &c., and stimulate the organs of secretion, so that those organs *work in harmony with each other.*

As the internal organs are all more or less dependent upon each other, and unite their functions for similar ends, it is obvious that there should prevail among them a *mutual consciousness* of their state. Otherwise, when the stomach had formed chyme, the liver might have no bile ready to fulfil its office; the absorbents might be in a state of rest at the moment when nutrition was set before them; and the heart might beat slowly, while the lungs were in active exertion to obtain additional blood to support an active exercise. The sympathetic system of nerves therefore *regulates and harmonises these internal functions.*

CHAPTER XLVIII.

963. *Why do we see objects?*

Because the light which is reflected from them enters our eyes and produces images of their forms upon a membrane of nerves called the *retina*, just as images are produced upon a mirror.

964. *Why does this enable us to see?*

Because the membrane which receives the images of objects is connected with the *optic nerve* which transmits to the brain impressions made by the reflections of light, just as other nerves convey the effects of feeling, hearing, tasting, &c.

965. *Why are we enabled to move our eyes?*

Because various muscles are so placed in relation to the eyeball, that their contraction draws the eye in the direction required. We are thus enabled to adjust the direction of the eye to the position of the objects we desire to see, in other words to *set the mirror in such a position that it will receive the reflection.* (See 517.)

"Truly the light is sweet, and a pleasant thing it is for the eyes to behold the sun."—ECCLESIASTES XI.

966. *Why are we enabled to see large objects upon so small a surface?*

Because the lenses and humours of the eye *collect the rays of light* coming from every direction, and, *bringing them into a focus*, transmit them to the retina, where each ray impresses upon the nervous surface the qualities it received from the object which reflected it.

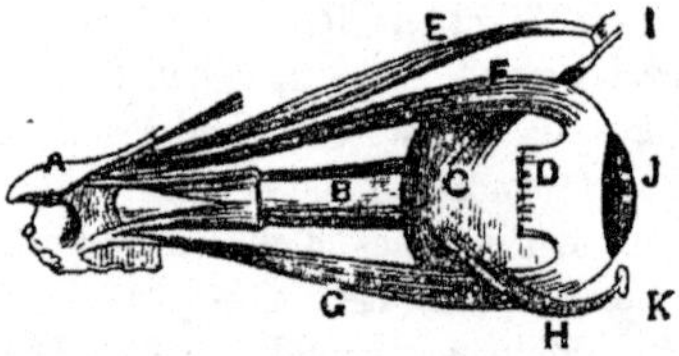

Fig. 61.—THE EYEBALL AND ITS MUSCLES.

A. Portion of *bone* through which the optic nerve passes in its communication between the brain and the eye.

B. The *optic nerve*, from before which an *external muscle* has been cut away, leaving its two attachments.

C. The *globe* of the eye.

D. The muscle which turns the eye *outward*, and which is counteracted by a muscle on the other side.

E. The muscle which passes through a loop, or staple of cartilage I, and *turns the eye obliquely*. It is counteracted by a muscle situated underneath.

F. The muscle situated underneath, which *turns the eyeball upwards*, and is counteracted by

G. The muscle which *turns the eyeball downwards*.

H. The muscle attached to a bone which *turns the eyeball upwards*.

I. The *cartilaginous loop* through which a muscle passes.

J. The front chamber of the eye filled with a clear fluid.

K. Fragment of the bone by which one of the muscles is fastened.

967. *Why do some persons squint?*

Because it sometimes happens that a muscle of the eye *acts too powerfully* for its companion muscle, and draws the eye too much on one side.

968. *Why does the pupil of the eye look black?*

Because the pupil is an *opening* through which the rays of light pass into the chamber of the eye. There is, therefore, nothing in the pupil, of the eye to reflect light.

"Keep me as the apple of thine eye; hide me under the shadow of thy wings"—PSALM XVII.

969. *Why is the pupil of the eye larger sometimes than at others?*

Because the *iris*, a ring of extremely fine muscles which surround the pupil, contracts when too much light falls upon the retina, and dilates when the light is feeble. It therefore enlarges or diminishes the size of the pupil to *regulate the admission of light.*

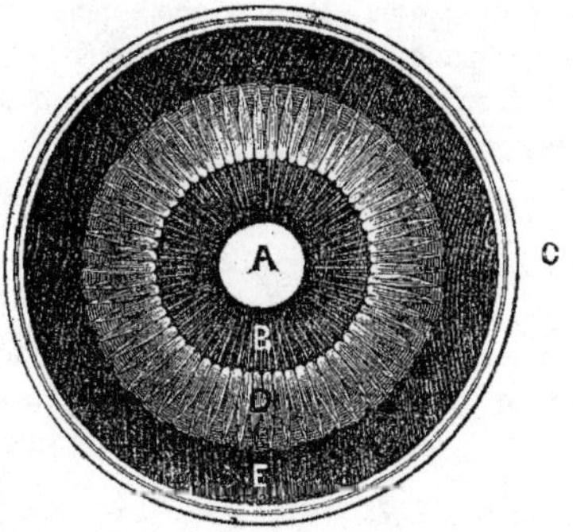

Fig. 62.—SECTION OF THE EYE SEEN FROM BEHIND

A. The *pupil* of the eye through which the light enters.

B. The *iris*, which dilates or contracts, and thereby increases or lessens the size of the pupil.

C C. The three coats of the eye, called the *sclerotic*, *choroid*, and *retina*.

D. The *ciliary processes*, or hair-like muscles, which have a slight vibratory motion which they impart to the fluids of the eye.

E. The dark coat of the *choroid*, the coat forming the *retina* removed.

970. *Why have we two eyes?*

Because the field of vision is thereby *much extended;* the *intensity* of sight is also increased, the impressions upon the brain being clearer and better defined, just as in a *stereoscope* the effect of vision is heightened by a double picture; the sense of sight being more *constantly* exercised than any other sense during our waking moments, *one eye is frequently called upon to give rest to the other;* and the important faculty of vision, being endangered by the necessary exposure of some parts of the eye, and the equally necessary delicacy of an organ formed to receive impressions from so ethereal an element as light, is rendered the more secure to us, since though one eye may become enfeebled, diseased, or wholly lost, *the other eye will retain the blessing of sight.*

971. *Why, having two eyes, and each eye receiving a reflection upon its retina, does the brain experience only one impression of an object?*

Because, besides those optical laws which bring upon the two retinas the exactly corresponding images of the same objects, the optic nerves *meet* before they reach the brain, *and blend the impulses which they convey.*

972. *Why are the eyes provided with eye-lids?*

Because the eyes require to be *defended* from floating particles in in the air, and to be kept *moist and clean.* The eye-lids form the shutters of the eye, defending it when waking, by closing upon its surface whenever danger is apprehended, moistening its surface when it becomes dry, and covering it securely during the hours of sleep.

973. *Why are the eye-lids fringed with eye-lashes?*

Because the eye-lashes assist to modify the light, and to protect the eye, without actually closing the eye-lids. When the eye-lids are partially closed, as in very sunny or dusty weather, the eye-lashes cross each other, forming a kind of shady lattice-work, from the interspaces of which the eye looks out with advantage, and sees sufficiently for the guidance of the body.

974. *Why are we able to see at long or short distances?*

Because the *crystalline lens* of the eye is a moveable body, and is pushed forward, or drawn back by fine muscular fibres, according to the distances of the objects upon which we look. By these means its *focus* becomes adjusted.

975. *Why do we wink?*

Because, by the repeated action of winking, *the eye is kept moist and clean,* and the watery fluid secreted by little glands in the eye-lids, and at the sides of the eye, is spread equally over the surface, instead of being allowed to accumulate. But the action of

"And the eye cannot say unto the hand, I have no need of thee; nor again the head to the feet, I have no need of you."—CORINTH. XII.

winking, or brightening the eye, is so instantaneous that it does not impede the sight.

976. *Whence are the humours and secretions of the eye derived?*

From the blood, which flows abundantly to the eyes, and is circulated in capillary vessels that are spread out upon the membranous coats of the eye-balls.

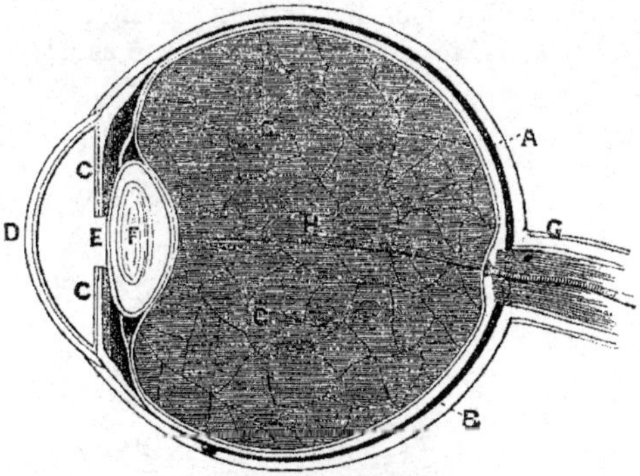

Fig. 63.—SECTION OF THE EYE.

A and B. The *sclerotic, choroid,* and *retina,* the three layers or coats which form the walls of the globe of the eye, and enclose its humours.
C C. The *iris.*
D. The front chamber of the eye, filled with watery humour.
E. The *pupil,* through which the rays of light pass to
F. The *crystalline lens.*
G G. The *vitreous humour* enclosed in cells formed by the *hyaloid membrane.*
H. An *artery* which supplies blood to the *crystalline lens,* and which passes through the centre of the *optic nerve.*
G. The *optic nerve,* showing the sheath in which the nerve is enclosed.

977. *Why do tears form in the eyes?*

Because, under the emotions of the mind, the circulation of blood in the brain, and in its nearest branches, becomes considerably quickened. The eyes receive a larger amount of blood, and the secretion of the lachrymal glands being increased, the fluid overflows, and tears are formed. The use of tears is probably *to keep the eyes cool during the excitement of the brain.* They are formed also during *laughing,* but less frequently.

If the whole body were an eye, where were hearing? if the whole were hearing where were smelling?"—CORINTHIANS XII.

978. *Why do we feel inconvenienced by sudden light?*

Because an excess of light enters the eye before the *iris* has had *time to adjust the pupil* to the amount of light to be received.

979. *Why if we look upon a very bright light, and then turn away, are we unable to see?*

Because the *iris* has so reduced the pupil while we were looking at the bright light, that immediately upon turning to a darker object, *the pupil is too small* to admit sufficient rays to enable us to see.

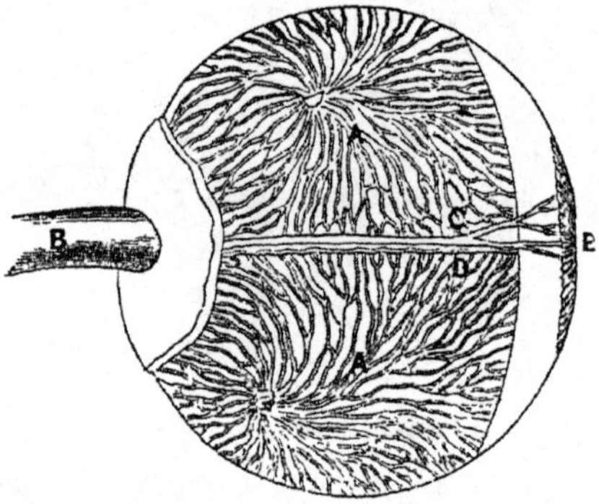

Fig. 64.—CAPILLARY BLOOD-VESSELS OF THE EYE.

A A. Capillary veins distributed over the *sclerotic coat.*

B. One of the trunks of the *optic nerve.*

C. A *nerve* communicating with the *ciliary processes.*

D. A *vein* running parallel with the nerve to the *ciliary processes.*

E. Side view of the *iris.*

980. *Why do we see better after a short time?*

Because the *iris* has relaxed and enlarged the pupil, therefore *we receive more rays of light* from the comparatively dark object, and are enabled to see it more clearly.

981. *Why do cats, bats, owls, &c., see in the dark?*

Because their eyes are made highly sensitive to *small quantities of light.* It is also believed that there are certain properties of light which affect their eyes, but do not affect ours. In other words, that there are some rays which are luminous to them which are not luminous to us. Hence they find *light* in what we call *darkness.*

982. *Why does the pupil of a cat's eye appear nearly closed by day?*

Because the cat's eye is so sensitive to light that the iris *closes the pupil almost entirely* to shut out the too powerful light.

CHAPTER XLIX.

983. *Why do we hear?*

Because the *tympanum* of the ear *receives impressions from sounds,* and transmits those impressions to the brain in a similar manner to that in which the retina of the eye transmits the impressions made upon it by light.

984. *Why is one part of the ear spread out externally?*

The external ear is a *natural ear-trumpet,* and serves to collect the vibrations of sound, and to conduct them towards the internal ear.

985. *Why is the ear allowed to project, whilst the eye is carefully enclosed?*

Because the external ear, being formed of tough cartilaginous substance, and being very simple in its organisation, is but little liable to injury.

986. *Why do hairs grow across the entrance of the ears?*

Because they prevent the intrusion of insects, and of particles of dust, by which otherwise the faculty of hearing would be impaired.

The insect called the *ear-wig* is popularly supposed to be so named from its tendency to get into the human ear, and cause pain and madness by penetrating to the brain. An earwig, however, is no more likely to get into the ear than any other insect whose habit it is to penetrate the corollas of flowers; and should an insect enter the ear, it could get no further than the *membrane of the tympanum,* which spreads all over the auditory passage, just as the parchment of a drum spreads over the entire circumference of that instrument. The fact is, that the wing of the insect, when spread, *resembles the external ear* in shape. It is similar to the wing of the stag beetle (*see* illustration), and this fancied resemblance of the wing of the insect to the ear of man may have given rise to the name of *ear-wing,* which became corrupted to *ear-wig.*

"Doth not the ear try words? and the mouth taste his meat."—JOB XII.

987. *Why is wax secreted at the entrance of the ear?*

Because, by the peculiar resinous property which it possesses, *it improves the sound-conducting power* of the auditory canal through which it prevails.

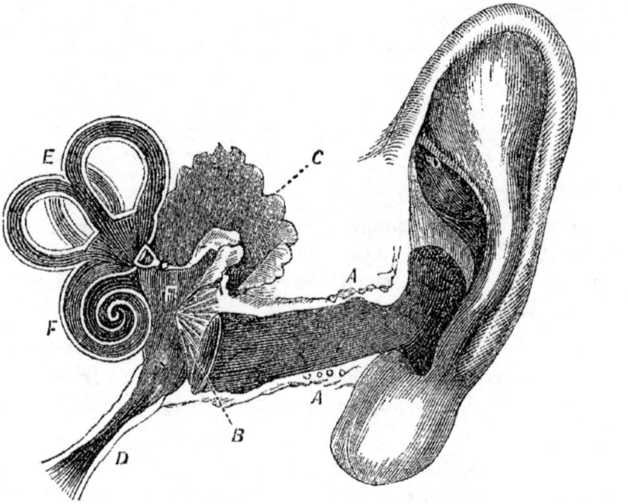

Fig. 65.—THE STRUCTURE OF THE EAR.

A A. Glands which secrete wax in the walls of the tube of the ear.

B. The membrane of the *tympanum*, or drum of the ear, formed in the shape of a funnel.

C C. Bones which act as a sort of sounding-board to the ear, giving strength to the vibrations.

D. The Eustachian tube, which opens into the root of the mouth, and which serves to preserve an equilibrium in the density of the air occupying the tubes of the ear.

E and F. The *labrynth* of the ear, consisting of folds of membraneous tubes, filled with fluid, which serves to undulate with the vibrations of the *tympanum*, and thus gives clearness and precision to the sounds.

The *auditory nerves* are distributed in the tubes above described (the *vestibule* and the *cochlea* E F), and the nerves receive their impressions from the undulations of the fluid.

988. *Why do we sometimes hear singing noises in the ear?*

Because the ear is liable to inflammation from various causes, and when the blood flows unduly through the vessels of the ear it *produces a slight sound.*

989. *Why do people become deaf?*

Because the ear may be injured in various ways: the tympanum may be impaired, the fluid of the ear dried up, or the nerves be pressed upon by swellings in the surrounding parts. When, therefore, the *mechanism of hearing* is impaired, the sense of hearing becomes weakened, or altogether lost.

990. *Why do persons accustomed to loud noises feel no inconvenience from them?*

Because the ***sensitiveness*** of the nerves of the ear becomes deadened. They do not convey to the brain such intense impulses when they are frequently acted upon by loud sounds.

991. *Why do persons engaged in battle often lose their hearing?*

Because the vibrations caused by the sounds of artillery are so violent that they overpower the mechanism of the ear, and frequently *rupture the connection of the fine. nervous filaments* with the textures through which they spread.

The violent concussions of the air produced by volleys of cannon, or by loud peals of thunder, have an overpowering effect upon persons nervously constituted, and upon the organ of hearing, which is more especially affected. As persons have been struck blind by intense light, so others have been deafened by intense sounds. In 1697 a butcher's dog was killed by the noise of the firing to celebrate the proclamation of peace. Two troops of horse were dismounted, and drawn up in a line to fire volleys. At the moment of the first volley a large and courageous mastiff, belonging to a butcher, was lying asleep before the fire. At the noise of the first volley the dog started up, and ran into another room, where it hid itself behind a bed; on the firing of the second volley, it ran several times bout the room, trembling violently; and when the hird volley was fired it ran around once or twice with great voilence, and then dropped down dead, with blood flowing from its mouth and nose. Persons who are painfully affected by loud noises should put a little wool in their ears when such noises are occurring; they will thereby save themselves from temporary inconvenience, and probably preserve the sense of hearing from permanent injury.

992. *Why do we smell?*

Because minute particles of matter, diffused in the air, come in contact with the filaments of the *olfactory nerve,* which are spread out upon the walls of the nostrils, and those nerves transmit impressions to the brain, constituting what we call the *odour of substances.*

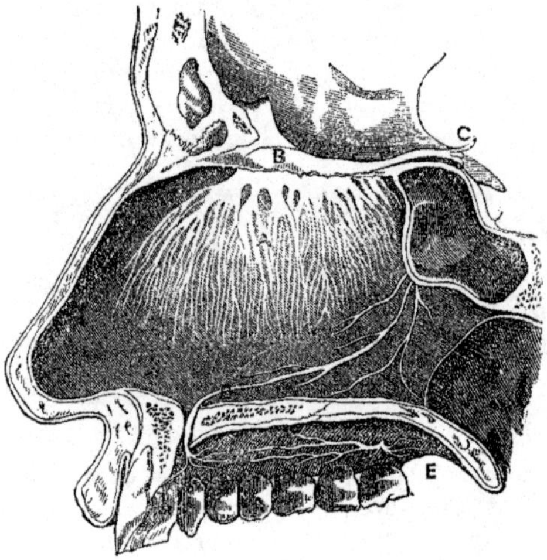

Fig. 66.—SHOWING THE DISTRIBUTION OF THE NERVOUS FILAMENTS UPON SENSITIVE MEMBRANES.

A. The *olfactory nerve,* distributed in minute branches upon the membrane of the nostril.

B. The *bulb* of the *olfactory* nerve.

C. The *roots* from which the olfactory nerve originates.

D E. *Nerves of the palate,* showing the manner in which they are passed through the bones of the roof of the mouth.

993. *Why do hairs grow across the passages of the nostrils?*

Because they form a *defence* against the admission of dust and insects, which would otherwise frequently irritate the nervous structure of the nose.

994. *Why are the nostrils directed downwards?*

Because, as odours and effluvia *ascend,* the nose is directed towards them, and thereby receives the readiest intimation of those bodies floating in the air which may be pleasurable to the sense, or offensive to the smell, and injurious to life.

995. *Why is the nose placed over and near the mouth?*

Because, as one of the chief duties of that organ is to *exercise a watchfulness* over the purity of the substances we eat and drink, it is placed in that position which enables it to discharge that duty with the greatest readiness.

CHAPTER L.

996. *Why do we taste?*

Because the tongue is endowed with *gustatory* nerves, having the function of *taste* as their *special sense,* just as the *optic,* the *auditory,* and the *olfactory* nerves, have their special duties in the eyes, ears, and nose.

997. *Why do some substances taste sweet, others sour, others salt, &c.?*

It is believed that the impressions of taste arise from the various *forms of the atoms of matter* presented to the nerves of the tongue.

998. *Why do we taste substances most satisfactorily after they have remained a little while in the mouth?*

Because the nerves of taste are most abundantly distributed to the under surface of the tongue; and when solid substances have been in the mouth a little while, they impregnate the saliva of the mouth with their particles *and come in contact in a fluid solution with the gustatory nerves.*

999. *Why if we put a nub of sugar to the tip of the tongue has it no taste?*

Because the gustatory nerves are *not distributed to that part of the tongue.*

"Wine is a mocker, strong drink is raging; and whosoever is deceived thereby is not wise."—PROVERBS XX.

1000. *Why, when we draw the tongue in, do we recognise the sweetness of the sugar?*

Because the dissolved particles of sugar are *brought in contact* with the nerves of taste.

1001. *Through what nerves are we made sensible of the contact of sugar with the tip of the tongue?*

Through the nerves of *feeling*, which are abundantly distributed to the tongue to guide it in its controul over the mastication of food.

1002. *Why do conniseurs of wines close their mouths and distend their chins for a few seconds, when tasting wines?*

Because they thereby bring the wine in contact with the under surface of the tongue, *in which the gustatory nerves chiefly reside.*

1003. *Why do they also pass the fumes of the wines through their nostrils?*

Because *flavour*, in its fullest sense, comprehends not only the *taste*, but the *odour* of a substance; and, therefore, persons of experience attend to both requisites.

The various conditions of taste are defined to be:—

1. Where sensations of *touch* are alone produced, as by glass, ice, pebbles, &c.

2. Where, in addition to being *felt* upon the tongue, the the substance excites sensation in the *olfactory nerves*, as by lead, tin, copper, &c.

3. Where, besides being *felt*, there are peculiar sensations of *taste*, expressive of the properties of bodies, as salt, sugar, tartaric acid, &c.

4. Where, besides being *felt* and *tasted*, there is an *odour* characteristic of the substance, and essential to the full developement of its flavours, as in cloves, lemon-peel, carriway-seed, and aromatic substances generally.

1004. *Why do we feel?*

Because there are distributed to various parts of the body fine nervous filaments, which have for their special duty the transmission to the brain of impressions made upon them *by contact* with substances.

1005. *In what parts of the body does the sense of touch more especially reside?*

In the points of the fingers and in the tongue. By laying a piece of paper upon a table, and upon the paper a piece of cloth, on the piece of cloth a bit of silk, and on the bit of silk a piece of leather, so that the edge of each would be exposed to the extent of half-an-inch, it would be possible by the touch to tell when the finger passed successively over the leather, silk, cloth, or paper, and arrived on the table.

Those impressions of touch must have been communicated, with their extremely nice distinctions, to the sensitive nerves that lie underneath the skin, and must have been transmitted all the way through the arm to the brain, although the touch itself was so light as scarcely to be appreciable with regard to the force applied.

A hair lying on the tongue will be plainly perceptible to the touch of the tongue; and the surface of a broken tooth will often causes the tongue great annoyance, by the acute perception it imparts of the roughness of its surface.

The toes are also highly sensitive, though their powers of touch are seldom fully developed. Persons who have lost their arms, however, have brought their feet to be almost as sensitive as fingers. Blind persons increase, by constant exercise, their powers of touch to such a degree that they are able to read freely by passing their fingers over embossed printing; and they have been known to distinguish *colours* by differences in their grain, quite unappreciable by other persons.

1006. *Why is feeling impaired when the hands are cold?*

Because, as the blood flows slowly to the nerves, they are less capable of that perception of touch which is their *special sense.* The skin contracts upon the nervous filaments, and *impairs the contact* between them and the bodies which they touch.

1007. *Why do the fingers prick and sting when they again become warm?*

Because, as the warmth expands the cuticle, and the blood begins to flow more freely through the vessels, *the nerves are made conscious of the movements of the blood,* and continue to be so until the circulation is equally restored to all the parts.

"In the sweat of thy face shalt thou eat bread, till thou return to the ground for out of it thou wast taken: for dust thou art, and unto dust shalt thou return."—GENESIS III.

1008. *Why do persons whose legs and arms have been amputated fancy they feel the toes or fingers of the amputated limb?*

Because the nervous trunk which formerly conveyed impressions from those extremities remains in the part of the limb attached to the body. *The mind has been accustomed to refer the impulses received through that nervous trunk to the extremity where the sensations arose.* And now that the nerve has been cut, the painful sensation caused thereby is referred to the extremity which the nerve supplied, and the sufferers for a time appear to *continue to feel the part which they have lost.*

CHAPTER LI.

1009. *Why do we perspire?*

Because the skin is filled with very *minute pores,* which act as outlets for a portion of the water of the blood, that serves to *moisten and cool* the surface of the body, and to carry away some of the matter *no longer needed in the system.*

1010. *How is the perspiration formed?*

By very small *glands,* which lie embedded in the skin. It is estimated that there are about 2,700,000 perspiratory glands distributed over the surface of the body, and that these glands find outlets for their secretion through no less than *seven millions of pores.*

1011. *What is insensible perspiration?*

Insensible perspiration is that *transmission of watery particles through the skin* which is constantly going on, but which takes place so gently that it cannot be perceived. It is, however, very important in its results, as no less than *from twenty to thirty-three ounces of water may pass imperceptibly through the skin in twenty-four hours.*

1012. *What is sensible perspiration?*

"And Elisha sent a message unto him, saying, Go and wash in Jordan seven times, and thy flesh shall come again to thee, and thou shalt be clean."—II KINGS V.

Sensible perspiration is that moisture which exudes upon the skin *in drops large enough to be perceptible,* when the body is heated by exercise or other means.

1013. *Why does a sudden change from heat to cold bring on illness?*

Because the effect of cold *arrests the action of the vessels of the skin,* and suddenly throws upon the internal organs the excretory labour which the skin should have sustained.

1014. *Why does a chill upon the skin frequently produce inflammation of the lungs?*

Because the lungs and the skin together discharge the chief proportion of the watery fluid of the body. *When the skin's action is checked, the lungs have to throw off a much greater amount of fluid.* The lungs, therefore, become *over worked,* and inflammatory action sets in.

1015. *Why does cleanliness promote health?*

Because every atom of dirt which lodges upon the surface of the body serves to clog and check the working of those *minute pores,* by which much of the fluid of the body is changed and purified.

In the internal parts of the system, the Creator has made ample provision for cleanliness. Every organ is so constituted that it cleanses and lubricates itself. Every surface of the inner body is perfectly clean, and as soft as silk

Nature leaves to man the care of those surfaces which are under his immediate observation and controul; and he who, from idleness, or indifference to nature's laws, is guilty of personal neglect, *opposes the evident intentions of the Creator,* and must sooner or later pay the penalty of disobedience.

1016. *Why does exercise promote health?*

Because it *assists all the functions upon which life depend.* It quickens the circulation, and thereby nourishes every part of the body, causing the bones to become firm, and the muscles to become full and healthy. It promotes breathing, by which oxygen is taken into the system, and carbon thrown off, and thereby it produces a higher degree of organic life and strength than would otherwise exist. It

"Love not sleep lest thou come to poverty: open thine eyes, and thou shalt be satisfied with bread."—PROV. XX.

promotes perspiration, by which, through the millions of pores of the skin, much of the fluid of the body is changed and purified. And it induces that genial and diffused warmth, which is one of the chief conditions of a high degree of vitality.

1017. *Why do we feel fatigue?*

Because those organs which stimulate the mechanism of the body to act, *themselves require rest and repair.* When the brain and nerves arrive at that state, they make their condition known to the system generally, by indications which we denominate *fatigue.*

1018. *Why, after rest, do we return invigorated to our labours?*

Because the nervous system has accumulated, during the hours of rest, a fresh amount of that *vital force* which we call the nervous fluid, and by which the various organs of the body are excited to perform the duties assigned to them.

1019. *What is sleep?*

Sleep is understood to be that state of the body in which *the relation of the brain to some parts of the body is temporarily suspended.*

There are some parts of the body that *never sleep:* such are the heart, the lungs, the organs of circulation, and those parts of the nervous system that direct their operations.

But when sleep overtakes the system, it seems as if the *relations* of those parts under the controul of the will were temporarily suspended; as if, for instance, those nerves which move the arms, the legs, the eyes, the tongue, &c., were all at once unfastened, just as the strings of an instrument are relaxed by the turning of a key, or the throwing down of a bridge over which they were stretched.

What is meant by the temporary suspension of the relation of the brain to some parts of the body, may be thus explained. Notice a man when he sits dosing in a chair: at first his head is held up, the brain controlling the muscles of the neck, and keeping the head erect. But drowsiness comes on, the brain begins to withdraw its influence, and the muscles of the neck becoming as it were "unstrung," the head drops down upon the breast. But the sleep is unsound, and disturbed by surrounding noises. The brain is therefore frequently excited to return its influence to the muscles, and draw up the head of the sleeper. He gives a sudden start, every muscle is tightened in an instant, up goes the head, the eyes open, the ears listen, until a feeling of security and composure returns; the sleep again deepens, *the nervous connection is again withdrawn*, and then down drops the head as before.

1020. *Why do we dream?*

Dreams appear to arise *from the excitement of the brain during those hours when its connection with the other parts of the living organism is suspended.* For instance: a man dreams that he is pursued by a furious animal, and the mind passes through all the excitement of flying from danger; but the *connection* between the moving power, and the machinery of motion being suspended, no motion takes place. The same impressions upon the brain, when the nerves were "strung" to the muscles, would have caused a rapid flight, and a vigorous effort to escape from the apprehended danger.

1021. *Why do suppers, when indigestible substances are eaten, produce dreaming?*

Probably because, as the digestive organs are oppressed, and those parts of the nervous system which stimulate the organs of digestion are *excited by excessive action*, those portions of the brain which are not immediately employed by the digestive process are disturbed by that *sympathy* which is observed to prevail between the relative parts and functions of the body.

1022. *Why do we yawn?*

Because, as we become weary, the nervous impulses which direct the respiratory movements are enfeebled. It has been said that those movements are involuntary, and that the parts engaged in producing them are not subject to fatigue. But the operation of breathing is, *to some extent, voluntary*, though when we cease to direct it voluntarily, it is involuntarily continued by organs which know no fatigue.

When, therefore, we feel weary—still controuling our breathing in our efforts to move or to speak—there frequently arrives a period when, for a few seconds, the respiratory process is suspended. It seems to be the point at which the voluntary nerves of respiration are about to deliver their office over to the involuntary nerves; but

"And it shall be, when they say unto thee, Wherefore sighest thou? that thou shalt answer, For the tidings, because it cometh; and every heart shall melt, and all hands shall be feeble."—EZEKIEL XXI.

the pause in the respiration has caused a momentary deficiency of breath, and the involuntary nerves of respiration, coming suddenly to the aid of the lungs, cause a spasmodic action of the parts involved, and *a yawn*, attended by a *deep inspiration to compensate for the cessation of breathing*, are the result.

1023. *Why do we cough?*

Because the respiratory organs are excited by the presence of some body foreign or unnatural to them. A cough is an effort on the part of the air tubes to free themselves from some source of irritation. And so important are the organs of breathing to the welfare of the body, that the muscles of the chest, back, and abdomen, *unite in the endeavour to get rid of the exciting substance.*

1024. *Why do we sneeze?*

Because particles of matter enter the nostrils and excite the nerves of feeling and of smell. In sneezing, as in coughing, the effort is to free the parts affected from the intrusion of some matters of an objectionable nature. And in this case, as in the former one, there is a very general sympathy of other organs with the part affected, and an energetic effort to get rid of the evil.

1025. *Why do we sigh?*

The action of sighing arises from very similar causes to those of yawning. But in sighing, the nervous depression is caused by *grief*; while in yawning, it is the result of *fatigue*. In sighing, the effect is generally caused by an *expiration*—in yawning by an *inspiration*. The mind, wearied and weakened by sorrow, omits for a few seconds to continue the respiratory process; and then suddenly there comes an involuntary expiration of the breath, causing a faint sound as it passes the organs of the voice.

1026. *Why do we laugh?*

Laughing is caused by the very opposite influences that produce sighing. The nervous system is highly excited by some external cause. The impression is so intense, and the mind so fixed upon it, that the respiratory process is irregular, and uncontrolled. Persons excited to a fit of laughter generally hold their breath until they can hold it no longer, and then suddenly there is a quick expiration causing eccentric sounds, the mind being too intently fixed upon the cause of excitement, *either to moderate the sounds, or to controul the breathing.*

1027. *Why do we hiccough?*

Hiccough is caused by a spasmodic twitching of the diaphragm, a thin muscular membrane which divides the chest from the abdomen. It generally arises from sympathy with the stomach; and it is highly probable that the muscular twitches and jerks are so many efforts on the part of the diaphragm to *assist the stomach to get rid of some undigested matter.*

1028. *Why do we snore?*

Snoring is caused by air sweeping through the passages that lead from the mouth through the nostrils, and which, in our waking moments, are capable of certain muscular modifications to adapt them to our breathing. But as in sleeping the nervous controul over them is withdrawn, they are left to the action of the air which, in sweeping by them, *sets them in vibration.*

We have endeavoured, by the employment of the simplest language, and by reference to some of the most familiar phenomena of nature, to impart to the reader a clear conception of those sublime laws which controul our being, and afford evidence of the goodness and power of that Almighty God to whom we are indebted for the life that we enjoy, and the varied and beautiful existences which, to the rightly constituted mind, make the earth a vast aggregation of interesting objects. We will now, before we pass on to the final section of our work, review some of the more important facts that have been communicated, and devote a few pages to meditations upon the formation of the human body—that wonderful temple of which each of us is a tenant.

We have described man's organisation. What is that organisation for? *It is to make use of the elements upon which man exists.* The lungs make use of the air; the eye makes use of the light; the stomach, and the system generally, make use of water; every part of the body uses heat; and all parts of the system demand food. The hand feeds as constantly as the mouth. The mouth is the receptacle of food, by which the body is to be fed; the stomach is the kitchen in which food is prepared for the use of the body; and the blood-vessels are the canals through which the food is sent to those members of the body that are in need of it. When we speak of man's "organs" or "members," we speak of those parts of the living machinery by which the elements are used up, or employed, for man's benefit. And this view of the subject, bearing in mind that the body is held together as the temple of a living Spirit, superior to mere flesh and blood, gives us a higher and clearer perception of the distinction between the body and the soul than that which we might otherwise entertain. The body is a machine, working for the spirit, which is its owner. While the machine works, the spirit directs and influences its actions. But when the machine stops, the spirit resigns its power over a ruined temple, quits it, and flies to a region where, as a spirit, it becomes subject to a new order of existence consistent with its severance from earthly things and laws, and there it enters upon its eternal destiny, according to the judgments and appointments of God. It is no longer dependent upon a relation between spiritual and material laws.

Suppose that the air which man breathes, instead of returning from his lungs clear and imperceptible to sight, were tinged with colour; we should see, that every time a man breathed, the air would rush in a stream into his mouth, and then return again; and the air which returned would, being warm, be lighter than the outer air, and would rise upward over the man's head, where, cooling and mingling with the outer air, it would descend again. We do, in fact, see this action evidenced; when in winter time the cold condenses the vapour of the breath, we see the little cloud constantly rising before the breather's face, and dispersing in the surrounding air.

Is it not a wonderful thing that that clear and elastic substance, which you cannot feel, though it touches every part of your body, and which you cannot see, is composed of two distinct bodies, having very different properties; and that the two bodies can easily be separated from each other?

Air is of the first importance to life. Hence it is provided for us everywhere. We require air every second, water every few hours, and food at intervals considerably apart. Air is therefore provided for us everywhere. Whether we stand or sit; whether we dwell in a valley or upon a mountain; whether we go into the cellar under our house, or into the garret at the top of it, air is there provided for us. God, who made it a law that man should breathe to live, also sent him air abundantly, that he might comply with that law. And all that is required from man in this respect is, that he will not shut out God's bounty, but receive it freely.

As we have employed the idea that if the air were coloured we should have the opportunity of marking the process of breathing, let us enlarge upon this, and suppose that every time the air were returned from the lungs it became of a darker colour, the darkness denoting increasing impurity. If we placed a man in a room full of pure air, we should see the air enter his lungs, and sent back slightly tinged; but this would disperse itself with the other air of the room and scarcely be perceptible. As the man continued to breathe, however, each measure of air returning from the lungs would serve to pollute that abiding in the room, until at last the whole mass would become cloudy and discoloured, and we should see such a change as occurs when water is turned from a pure and clear state into a muddy condition. The air does become polluted with each respiration, and although it is colourless, it is as impure as if with every breath given off from the lungs it became of a dark colour in proportion to its impurity.

Thus we see how important it is that we should provide ourselves with pure air; and that, in seeking warmth and comfort in our houses, we should provide an adequate supply of fresh atmosphere—because it is more vital to life than either water or food.

Indeed, so constant is our requirement of air, that *if we had to fetch it, for purposes of breathing, or simply to raise it to our mouths, as we do water when we drink, it would be the sole occupation of our lives—we could do nothing else.* For this reason, God has sent the air to us, and not required us to go to the air. And the great error of man is, that in too many instances, he shuts off the supply from himself, and brings on disease and pain by inhaling a poisonous compound, instead of air of a healthful kind, which bears an adaptation to the wants of life.

Whilst the rooms of our house are filled with air, it is otherwise with water, which we require in less degree than air. If we have not the artificia means by which water is brought to our houses, through the pipes of a water company, there is a spring or a pump in the garden; or in the absence of these a good sound cask, standing at the end of our house, forming a receptacle to the water-pipes that surround it, provides us with a supply of water distilled from the clouds. If we were to drink a good draught of water once a day, that would be sufficient for all the purposes of life, as far as regards the alimentary uses of water. Man is, therefore, allowed to go to the stream for his drink, and is required to raise it to his lips at those moments when he uses it.

Although, in breathing, man separates the *oxygen* of the air from the *nitrogen* thereof, he does not separate the *oxygen* of the water from the *hydrogen* Water, in fact, undergoes no change in the body, excepting that of admixture with the substances of the body. And its uses are, to moisten, to cool, to cleanse, and also to nourish the parts with which it comes in contact. But it affords no nourishment of itself, it mixes with the blood, of which it forms a material part, and is the means of conveying *the nourishment of the blood* to every part of the system. After it has filled this office, and taken up impurities that required to be removed, it is cast out of the system again, without undergoing any chemical change.

Man's body is to his Soul, in many respects, what a house is to its occupant. But how superior is the dwelling which God erected, to that which man has built. Reader, come out of yourself, and in imagination realise the abstraction of the Soul from the body. Make an effort of thought, and do not relinquish that effort, until you fancy that you see your image seated on a chair before you. And now proceed to ask yourself certain questions respecting your bodily tenement—questions which, perchance, have never occurred to you before; but which will impress themselves the more forcibly upon you, in proportion as you realise for a moment the idea of your Soul examining the body which it inhabits. There sits before you a form of exquisite proportions, with reference to the mode of life it has to pursue—the wants of the Soul for which it has to care, and which it has to guard, under the direction of that Soul, its owner and master.

Over the brows that mark the intellectual front of that fine form, there fall the auburn locks of youth, or the grey hair of venerable age. Each of those hairs is curiously organised. If you take a branch of a tree, and cut it across, you will find curious markings caused by vessels of various structure, all necessary to the existence of the plant. In the centre will be found either a hollow tube, or a space occupied by a soft substance called pith. Each hair of your head is as curiously formed as the branch of a tree, and in a manner not dissimilar, though its parts are so minute that the unaided eye cannot discern them. Every hair has a root, just as a tree has, and through this root it receives its nourishment. As the vessell which feed a plant are always proportionate to the size of the plant itself, how fine must be those vessels which form the roots of the hair, being in proportion to the size of the hair, which is in itself so small that the eye cannot see its structure? The hair is, in fact, an animal plant, growing upon the body in much the same manner that plants grow upon the surface of the earth. But how does this hair grow? Not alone by the addition of matter at its roots, pushing up and enlongating its stem: nourishment passes up through its whole length, and is deposited upon its end, just as the nourishment of a tree is deposited upon its extreme branches. If, after having your hair cut, you were to examine its ends by the microscope, you would discover the abrupt termination left by the scissors. But allow the hair to grow, and then examine it, and you will discover that it grows from its point which, in comparison with its former state, is perfect and fine. The reason why the beard is so hard is, that the ends of the hair are continually being shaved off. The hair of the beard, if allowed to grow, would become almost as soft as the hair of the head.

But why is man's head thus covered with hair? For precisely the same reason that a house is thatched—to keep the inmates warm. We might add, also, to give beauty to the edifice. But as beauty is a conventional quality—and if men were without it they would consider themselves quite as handsome as they do now—we will not enlarge upon the argument. Our bald-headed friends, too, might have reason to complain of such a partial hypothesis. The brain is the great organ upon which the health, the welfare, and the happiness of the system depends. The skull, therefore, may be regarded as analagous to the "strong box," the iron chest in which the merchant keeps his treasure. There is no point at which the brain can be touched to its injury, without first doing violence to the skull. Even the spinal cord runs down the back through a tunnel or tube, formed in a number of strong bones, so closely and firmly jointed together, that they are commonly termed "the back-bone."

Look at the eyebrows. What purpose do they fulfil? Precisely that of a shed, or arch placed over a window to shelter it from rain. But for the eyebrows the perspiration would frequently run from the brow into the eyes, and obscure the sight; a man walking in a shower of rain would scarcely be able to see; and a mariner in a storm would find a double difficulty in braving the tempest.

Now we come to the eye, which is the window of the Soul's abode. And what a window! how curiously constructed! how wisely guarded! In the eyelashes, as well as the eyebrows, we see the hair fulfilling a useful purpose, differing from any already described. The eyelashes serve to keep cold winds, dust, and too bright sun, from injuring or entering the windows of the body. When we walk against the east wind, we bring the tips of our eyelashes together, and in that way exclude the cold air from the surface of the eye; and in the same manner we exclude the dust and modify the light. The eyelashes, therefore, are like so many sentries, constantly moving to and fro, protecting a most important organ from injury. The eyelids are the shutters by which the windows are opened and closed. But they also cleanse the eye, keeping it bright and moist. There are, moreover, in the lids of each eye or window, little glands, or springs, by which a clear fluid is formed and supplied for cleansing the eye. The eye is placed in a socket of the skull, in which it has free motion, turning right or left, up or down, to serve the purpose of the

"Thou art of purer eyes than to behold evil, and canst not look on iniquity."
HABAKKUK I.

inhabitant of the dwelling. Of the structure of the eye itself we will not say much, for the engravings will afford a clearer understanding than a lengthy written description. But we would have you examine the formation of the iris of the living eye, the ring which surrounds the pupil. Hold a light to it, and you will find that the iris will contract and diminish the pupil; withdraw the light, and the iris will relax, and the pupil expand, thus regulating the amount of light. The images of external objects are formed upon the retina of the eye, a thin membrane, spread out upon the extremity of a large nerve, which proceeds immediately to the brain, and forms the telegraphic cord by which information is given to the mind, of everything visible going on within the range of sight.

Now, think for a few moments upon the wonderful structure of those windows of the body. Can you fancy, in the walls of your house, a window which protects itself, cleanses itself, and turns in any direction at the mere will of the tenant; and when that tenant is oppressed by excess of light, draws its own curtain, and gives him ease; and when he falls asleep, closes its own shutters, and protects itself from the cold and dust of night, and the instant he awakes in the morning, opens, cleanses itself with a fluid which it has prepared during the night, and kept in readiness; and repeats this routine of duty day after day for half a century, without becoming impaired? Such, nevertheless, is the wonderful structure of the window of the body—the eye.

In some scientific works that have recently been published, curious investigations have been made known. It has been shown that the eye is impressed momentarily, as a photographic plate is impressed by the rays of the sun. But the photography of the eye has this extraordinary quality—that one image passes away, and another takes its place immediately, without confusion or indistinctness. But the most wonderful assertion of all is, that under the excitement of memory these photographic images are restored, and that when, "in our mind's eye," we see the image of some dear departed friend, the retina really revives an image which once fell upon its sensitive surface, and which image has been stored up for many years in the sacred portfolio of its affections!

Another extraordinary assertion is one which comes supported by a degree of authencity that entitles it to consideration. It is said that the eye of a dead man retains an impression of the last picture that fell upon the faithful retina. Dr. Sandford, of America, examined the eye of a man named Beardley, who had been murdered at Auburn, and he published in the *Boston Atlas* the following statement:—"At first we suggested the saturation of the eye in a weak solution of atrophine, which evidently produced an enlarged state of the pupil. On observing this, we touched the end of the optic nerve with the extract, when the eye instantly became protuberant. We now applied a powerful lens, and discovered in the pupil, the rude, worn-away figure of a man, with a light coat, beside whom was a round stone, standing or suspended in the air, with a small handle, stuck in the earth. The remainder was debris, evidently lost from the destruction of the optic, and its separation from the mother brain. Had we performed the operation when the eye was entire in the socket, with all its powerful connection with the brain, there is not the least doubt but that we should have detected the last idea and impression made on the mind and eye of the unfortunate man. The picture would evidently be entire; and perhaps we should have had the contour, or better still, the exact figure of the murderer

"Be not rash with thy mouth, and let not thine heart be hasty to utter anything before God: for God is in heaven, and thou upon earth; therefore let thy words be few."—ECCLESIASTES V.

The last impression on the brain before death is always more terrible from fear than any other cause, and figures impressed on the pupil more distinct, which we attribute to the largeness of the optic nerve, and its free communication with the brain." Whether the supposition, which seems to be supported by the experiment above detailed, be correct or not, it is in no sense more wonderful than the facts which are already known respecting this curious and perfect organ.

The nose is given us for two purposes—to enable us to respire and to smell. As odours arise from the surface of the earth, the cup or funnel of the nose is turned down to meet them. In the nostrils hair again, serves a useful purpose. It not only warms the air which enters the nostrils, but it springs out from all sides, and forms an intersecting net, closing the nostrils against dust, and the intrusion of small insects. If by any means, as when taking a sharp sniff, foreign matters enter the nostrils, the nose is armed with a set of nerves which communicate the fact to certain muscles, and the organs of respiration unite with those muscles to expel the intruding substances. In this action, the diaphragm, or the muscle which divides the abdomen from the chest, is pressed down, the lungs are filled with air, the passage by which that air would otherwise escape through the mouth, is closed up, and then, all at once, with considerable force, the air is pressed through the nostrils, to free them from the annoying substance. So great is the force with which this action takes place, that the passage into the mouth is generally pushed open occasioning the person in whom the action takes place, to cry "'tsha!" and thus is formed what is termed a sneeze. As with the eye, so with the nose—innumerable nerves are distributed over the lining membrane, and these nerves are connected with larger nerves passing to the brain, through which everything relating to the sense of smell is communicated.

The nose acts like a custom-house officer to the system. It is highly sensitive to the odour of most poisonous substances. It readily detects hemlock, henbane, monk's hood, and the plants containing prussic acid. It recognises the fœted smell of drains, and warns us not to breathe the polluted air. The nose is so sensitive, that air containing a 200,000th part of bromine vapour will instantly be detected by it. It will recognise the 1,300,000th part of a grain of otto of roses, or the 13,000,000th part of a grain of musk! It tells us in the mornings that our bed-rooms are impure; it catches the first fragrance of the morning air, and conveys to us the invitation of the flowers to go forth into the fields, and inhale their sweet breath. To be "led by the nose," has hitherto been used as a phrase of reproach. But to have a good nose, and to follow its guidance, is one of the safest and shortest ways to the enjoyment of health.

The mouth answers the fourfold purpose of the organ of taste, of sound, of mastication, and of breathing. In all of these operations, except in breathing, the various parts of the mouth are engaged. In eating we use the lips, the tongue, and the teeth. The teeth serve the purpose of grinding the food, the tongue turns it during the process of grinding, and delivers it up to the throat for the purposes of the stomach, when sufficiently masticated. The lips serve to confine the food in the mouth, and assist in swallowing it, and there are glands underneath the tongue, and in the sides of the mouth, which pour in a fluid to moisten the food. And so watchful are those glands of their duty, that the mere imagination frequently causes them to act. Their fluid is required to modify

"I say unto you, Swear not at all; neither by heaven, for it is God's throne; Nor by the earth; for it is his footstool."—MATTHEW V.

the intensity of different flavours and condiments in which man, with his love of eating, will indulge. Thus, when we eat anything very acid, as a lemon, or anything very irritating, as Cayenne pepper, the effect thereof upon the sensitive nerves of the tongue is greatly modified by a free flow of saliva into the mouth. And if we merely fancy the taste of any such things, those glands are so watchful, that they will immediately pour out their fluid to mitigate the supposed effect.

In speaking, we use the lips, the teeth, the tongue; and the chest supplies air, which, being controlled in its emission, by a delicate apparatus at the mouth of the wind-pipe, causes the various sounds which we have arranged into speech, and by which, under certain laws, we are enabled to understand each other's wants, participate in each other's emotions, express our loves, our hopes, our fears, and glean those facts, the accumulation of which constitutes knowledge, enhances the happiness of man, and elevates him, in its ultimate results above the lower creatures to which the blessing of speech is denied.

The curious structure of the tongue, and the organs of speech, would fill a very interesting volume. The tongue is unfortunately much abused, not only by those who utter foul words, and convert the blessing of speech, which should improve and refine, into a source of wicked and profane language; but it constantly remonstrates against the abuse of food, and the use of things which are not only unnecessary for the good of our bodies, but prejudicial to their health. When the body is sufficiently fed, the tongue ceases its relish, and derives no more satisfaction from eating: but man contrives a variety of inventions to whip the tongue up to an unnatural performance of its duty, and thus we not only over-eat, but eat things that have no more business in our stomachs, than have the stones that we walk upon. Can we wonder, then, that disease is so prevalent, and that death calls for many of us so soon.

That wonderful essence, the Soul of man, rises above all finite knowledge. Its wonders and powers will never, probably, be understood until when, in a future state of existence, the grandest of all mysteries shall be explained. When we talk of the brain, we speak of that which it is easy to comprehend as the organ, or the seat of the mind; when we speak of the mind, we have greater difficulty in comprehending the meaning of the term we employ; but when we speak of the Soul, we have reached a point which defies our understanding, because our knowledge is limited. The brain may be injured by a blow; the mind may be pained by a disagreeable sight, or offended by a harsh word; but the Soul can only be influenced secondarily through the mind, which is primarily affected by the organs of the material senses. Thus the happiness or the misery of the Soul depends to a very great extent upon the proper fulfilment of the duties of the senses, which are the servants of the Soul, over which the mind presides, as the steward who mediates between the employer and the employed.

The Ear, which is taught to delight in sweet sounds, and in pure language, is a better servant of the master Soul, than one which delights not in music, and which listens, with approbation or indifference, to the oaths of the profane. The Eye which rejoices in the beauties of nature, and in scenes of domestic happiness and love, is a more faithful servant than one that delights in witnessing scenes of revelry, dissipation, and strife. The Nose which esteems the sweet odour of flowers, or the life-giving freshness of the pure air, is more dutiful to his master than one that rejects not the polluted atmosphere of

"Out of the same mouth proceedeth blessing and cursing. My brethren, these things ought not so to be."—JAMES III.

neglected dwellings. The Mouth which thirsts for morbid gratification of taste, is more worthless than one which is contented with wholesome viands, and ruled by the proper instincts of its duty. Who that can understand the wonderful structure of the tongue, and the complicated mechanism of the organs of speech and of hearing, could be found to take pleasure in the utterance of oaths, and of words of vulgar meaning? Were those beautiful cords that like threads of silk are woven into the muscular texture of the mouth, and along which the essence of life travels with the quickness of thought, to do the bidding of the will—were they given for no higher use than to sin against the God who gave them, and upon whose mercy their existence every moment depends?

The actions of the senses must necessarily affect the mind, which is the head steward of the Soul; and the Soul becomes rich in goodness, or poor in sin, in proportion as the stewardship, held by his many servants, is rightly or wrongfully fulfilled. As in an establishment where the servants are not properly directed and ruled, they often gain the ascendancy, and the master has no power over them, so with man, when he gives himself up to sensual indulgences. The Soul becomes the slave of the senses—the master is controlled by the servants.

With regard to the mechanism of motion, let us take the case of a man who is walking a crowded thoroughfare, and we shall see how active are all the servants of the Soul, under the influence of the mind. He walks along in a given direction. But for the act of volition in the mind, not a muscle would stir. The eye is watching his footsteps. There is a stone in his path, the eye informs the mind, the mind communicates with the brain, and the nerves stimulate the muscles of the leg to lift the foot a little higher, or turn it on one side, and the stone is avoided The eye alights on a familiar face, and the mind remembers that the eye has seen that face before. The man goes on thinking of the circumstance under which he saw that person, and partially forgets his walk, and the direction of his steps. But the nerves of volition and motion unite to keep the muscles up to their work, and he walks on without having occasion to think continually, "I must continue walking." He has not to make an effort to lift his leg along between each interval of meditation: he walks and meditates the while. Presently a danger approaches him from behind. The eye sees it not—knows no more, in fact, than if it were dead. But the ear sounds the alarm, tells the man, by the rumbling of a wheel, and the tramp of horses' feet, that he is in danger; and then the nerves, putting forth their utmost strength, whip the muscles up to the quick performance of their duty; the man steps out of the way of danger, and is saved. He draws near to a sewer, which is vomiting forth its poisonous exhalations. The eye is again unconscious—it cannot see the poison lurking in the air. The ear, too, is helpless; it cannot bear witness to the presence of that enemy to life. But the nose detects the noxious agent, and then the eye points out the direction of the sewer, and guides his footsteps to a path where he may escape the injurious consequences. A clock strikes, the ear informs him that it is the hour of an appointment; the nerves stimulate the muscles again, and he is hastened onward. He does not know the residence of his friend, but his tongue asks for him, and his ear makes known the reply. He reaches the spot—sits—rests. The action of the muscles is stayed; the nerves are for a time time at rest. The blood which had flown freely to feed the muscles while they were working,

"I am but a little child: I know not how to go out or come in."—I KINGS III.

goes more steadily through the arteries and veins, and the lungs, which had been purifying the blood in its course, partake of the temporary rest.

Let us remember that there are two sets of muscles, acting in unison with each other, to produce the various motions, they are known by the general terms of *flexors* and *extensors;* the first enable us to bend the limbs, the other to bring the limbs back to their former position. The flexors enable us to close the hand, the extensors to open it again. The flexors enable us to raise the foot from the ground, the extensors set the foot down again in the place desired. Consider for a moment the nicety with which the powers of these muscles must be balanced, and the harmony which must subsist between them in their various operations. When we are closing the hand, if the extensor muscles did not gradually yield to the flexors—if they gave up their hold all at once, the hand, instead of closing with gentleness and ease, would be jerked together in a sudden and most uncomfortable manner. If, in such a case, you were to lay your hand with its back upon the table, and wish to close the hand, the fingers would fall down upon the palm suddenly, like the lid of a box Again, consider how awkward it would be in such a case; our walk through the streets would become a series of jumps and jerks; when a man had raised his foot, after it had been jerked up, there it would stand fixed for a second before the opposite muscles could put on their power to draw it down again. This case is not at all suppositious· there is a derangement frequently observed in horses, in which one set of muscles becomes injured, and we may see horses suffering from this ailment, trotting along with one of their legs jerking up much higher than the others, and set down again with difficulty, just in the manner described.

It is also to be observed that very nice proportions must exist between the sizes of the muscles and the sizes of the bones. If this were not the case, our motions, instead of being firm and steady, would be all shaky and uncertain. In old persons the muscles become weak and relaxed; hence there is a tendency in the movements of the aged to fall, as it were, together; the head is no longer erect, the body bends, the knees totter, and the arms lean towards the body as for support.

In the child a somewhat similar state of things exists. The muscles have not been properly developed, nor have they been brought sufficiently under the controul of the nervous system. The child, therefore, totters and tumbles about, and it is not until it has stumbled and tumbled some hundreds of times in its little history, that the muscles have become strong enough to fulfil their office, or have been brought sufficiently under the controul of the nervous system, to perform well the various duties required from them.

In all these things, we recognise the perfection of the divine works. We are apt, too apt, to overlook this perfection, because it prevails in everything; but by speculating upon what inconveniences we might suffer, were not things ordained as they are, we obtain most convincing evidences of divine goodness and wisdom.

Having taken this view of the muscular system of the external man, let us turn our attention to the muscles of the internal organs. The muscles of which we have been speaking are called the voluntary muscles, because we have them under our own controul—they are subject to the influences of our will. But there is the other set of muscles. What are they? We talk of the beating, or of

"Watchman, what of the night? The watchman said, The morning cometh, and also the night: if ye will enquire, enquire ye; return, come."—ISAIAH XXI.

the palpitation, of the heart. But, what is it that causes the heart to beat? You cannot, if you wish it, make your heart beat more quickly or more slowly. Place your finger upon your pulse, and notice the degree of rapidity with which its pulsations follow. Now think that you should like to double the frequency of those pulsations. Say to the heart, with your inner voice, that you wish it to beat 120 times in a minute, instead of 60. It does not obey you; it does not appreciate your command. Now place your finger on the table, and your watch by the side of your hand, and tell your finger to beat 60 times in the minute, or 100 times, or 150 times, or 200 times, and the finger will obey you—because it is *moved by muscles which are subject to the will*, while the heart is composed of muscles which are *not subject to the will*. Why should this be? Why should man have the power to regulate his finger, and not to regulate his heart?

For the sustentation of our bodies it is needful that the blood should ever be in circulation. If the heart were to cease beating only for three or four minutes (perhaps less) life would be extinct. In this short time the whole framework of man, beautiful in its proportions, perfect in its parts, would pass into the state of dead matter, and would simply wait the decay that follows death. The eye would become dull and glazed, the lips would turn blue, the skin would acquire the coldness of clay—love, hope, joy, would all cease. The sweetest, the fondest ties would be broken. Flowers might bloom, and yield their fragrance, but they would be neither seen nor smelt; the sun might rise in its brightest splendour, yet the eye would not be sensitive to its rays; the rosy-cheeked child might climb the paternal knee; but there, stiff, cold, without joy, or pain, or emotion of any kind, unconscious as a block of marble, would sit the man *whose heart for a few moments had ceased to beat.*

How wise, then, and how good of God, that he has not placed this vital organ under our own care! How sudden would be our bereavements—how frequent our deaths, how sleepless our nights, and how anxious our days, if we had to keep our own hearts at work, and death the penalty of neglect.

And yet, before we were born, until we reach life's latest moment—through days of toil, and nights of rest—even in the moments of our deepest sin against the God who at the time is sustaining us, our hearts beat on, never stopping, never wearying, never asking rest.

This brings us to another reflection. Our arms get weary, our legs falter from fatigue, the mind itself becomes overtaxed, and all our senses fall to sleep. The eye sees not, the ear is deaf to sound, the sentinels that surround the body, the nerves of touch, are all asleep—you may place your hand upon the brow of the sleeping man, and he feels it not. Yet, unseen, unheard, without perceptible motion, or the slightest jar to mar the rest of the sleeper, the heart beats on, and on, and on. As his sleep deepens, the heart slackens its speed, that his rest may be the more sound. He has slept for eight hours, and the time approaches for his awakening. But is the heart weary—that heart which has toiled through the long and sluggard night? No! The moment the waking sleeper moves his arm, the heart is aware that a motion has been made, that effort and exercise are about to begin. The nerves are all arousing to action; the eyes turn in their sockets, the head moves upon the neck; the sleeper leaves his couch, and the legs are once more called upon to bear the weight of the body. Blood is the food of the eye, the food of the ear, of the foot, the hand, and every member of the frame. While they labour they must be fed—that is the condition of their life, the source of their strength. The heart, therefore, so far from seeking rest, is all fresh and vigorous for the labours of the day, and proceeds to discharge its duty so willingly, that we do not even know of the movements that are going on within us.

Thus we have seen the difference between the voluntary and the involuntary muscles, and we have perceived the goodness of our Creator in not entrusting to our keeping the controul of an organ so vital to life, as the heart.

But the heart is not the only organ which thus works unseen and unfelt. There are the lungs and the muscles of the chest, the stomach, and other parts occupying the abdomen, together with all those muscular filaments which enter into the structure of the coats and valves of the blood-vessels, and which assist to propel the blood through the system. All these are at work at every moment of man's life; and yet, so perfect is this complicated machinery, that we really do not know, except by theory, what is going on within us.

During the time that the sleeper has been at rest, the stomach has been at work digesting the food which was last eaten. Then the stomach has passed the macerated food into the alimentary canal, the liver has poured out its secretion, and produced certain changes in the condition of the dissolved food: and the lacteals, of which there may be many thousands, perhaps millions, have been busy sucking up those portions of the food which they knew to be useful to the system, whilst they have rejected all those useless and noxious matters upon which the liver, like an officer of health, had set his mark, as unfitting for the public use. This busy life has gone on uninterruptedly; every member of that body, every worker in that wonderful factory, has been unremitting in his duty, and yet the owner, the master, has been asleep, and wakes up finding every bodily want supplied!

Notwithstanding that much has already been said of the wonders that pertain to the eye, it has not yet been considered as the seat of *tears*, those mute but eloquent utterers of the sorrows of the heart. Beautiful Tear! whether lingering upon the brink of the eyelid, or darting down the furrows of the care-worn cheek—thou art sublime in thy simplicity—great, because of thy modesty—strong, from thy very weakness. Offspring of sorrow! who will not own thy claim to sympathy? who can resist thy eloquence? who can deny mercy when thou pleadest?

Every tear represents some in-dwelling sorrow preying upon the mind and destroying its peace. The tear comes forth to declare the inward struggle, and to plead a truce against further strife. How meet that the eye should be the seat of tears—where they cannot occur unobserved, but, blending with the beauty of the eye itself, must command attention and sympathy!

Whenever we behold a tear, let our kindliest sympathies awake—let it have a sacred claim upon all that we can do to succour and comfort under affliction. What rivers of tears have flown, excited by the cruel and perverse ways of man! War has spread its carnage and desolation, and the eyes of widows and orphans have been suffused with tears! Intemperance has blighted the homes of millions, and weeping and wailing have been incessant! A thousand other evils which we may conquer have given birth to tears enough to constitute a flood—a great tide of grief. Suppose we prize this little philosophy, *and each one determine never to excite a tear in another*. Watching the eye as the telegraph of the mind within, let us observe it with anxious regard; and whether we are moved to complaint by the existence of supposed or real wrongs, let the indication of the coming tear be held as a sacred truce to unkindly feeling, and our efforts be devoted to the substitution of smiles for tears!

There is only one other matter to which we think it necessary to allude, before we pass to the concluding section of our work. It has been said (162), that snow which is *white*, keeps the earth warm; that *white* as a colour is *cool*, and that *black* absorbs *heat* (230). These assertions may appear to be contradictory, and, taken in connection with the fact of the blackness of the skin of negroes in hot climates, may at a first glance be considered unsatisfactory. They are, however, perfectly reconcileable, and that too, without the slightest evasion of the real bearing of the asserted facts. White snow is warm *on account of its texture*, which, being woolly, forms a layer of non-conducting substance over the surface of the earth, and *keeps in its warmth;* white clothing, worn as a garment consisting of a thin material, is cool, because *the white colour* turns back the rays of the sun that fall upon it. Swansdown, although white, being a non-conductor, would be warm, because, though it would reflect the light and heat, it would confine and accumulate the heat of the body. The black skin of the negro is a *living texture*, and is not subject to the same laws that govern dead matter. The skin of the negro is largely provided with cells which secrete a fatty matter that acts as a non-conductor of the *external* heat, and also a much larger number of perspiratory glands than exist in the skins of Europeans. The perspiration cools the blood, and carries off the *internal* heat, while the oily matter gives a shining surface to the skin, and reflects the heat, to which the fatty matter presents itself as a non conducter. We see, therefore, that there are two express provisions for the cooling of the negroes' skin, independent of the colour. The skin of the Esquimaux who inhabits a cold country is *white*, though it might be supposed that a black skin would best conduce to the warmth of his body. But the Esquimaux has, underneath his skin, *a thick coating of fat*, by which the *internal heat* of the body is prevented from escaping.

This *resume* of the subjects embodied in the form of question and answer in the previous pages, will serve to impress the more important truths upon the mind of the reader, while it has enabled us to fill up many omissions necessitated by the arbitrary form of catechetical composition.

"Ask now the beasts, and they shall teach thee; and the fowls of the air, and they shall tell thee."—JOB XII.

CHAPTER LII.

1029. *Why are there so many bodily forms in the animal creation?*

Because the various creatures which God has created have different modes of life, and the forms of their bodies will be found to present *a perfect adaptation to the lives allotted to them.*

Because, also, the beauty of creation *depends upon the variety of objects of which it consists.* And the greatness of the Creator's power is shown *by the diversity of ends accomplished by different means.*

1030. *Why are birds covered with feathers?*

Because they require a high degree of *warmth*, on account of the activity of their muscles; but in providing that warmth it was necessary that their coats should be of the *lightest material*, so as not to impair their powers of flight; and feathers combine the *highest warming power, with the least amount of weight.*

1031. *Why have ostriches small wings?*

Because, having long legs, they do not require their wings for flight; they are merely used *to steady their bodies while running.*

1032. *Why are ostrich feathers soft and downy?*

Because, as the feathers are not employed for flight, the *strength of the feather as constructed for flying is unnecessary*, and the feathers therefore consist chiefly of a soft down.

1033. *Why have water-birds feathers of a close and smooth texture?*

Because such feathers keep the body of the bird warm and dry, by repelling the water from their surface. A bird could scarcely move through the water, with the downy feathers of the ostrich, because of *the amount of water the down would absorb.*

1034. *Why is man born without a covering?*

Because *man is the only animal that can clothe itself.* As in the various pursuits of life he wanders to every part of the globe, he can adapt himself *to all climates and to any season.*

1035. *Why do the furs of animals become thicker in the winter than in the summer?*

Because the creator has thus provided for the preservation of the warmth of the animals during the cold months of winter.

1036. *Why does a black down grow under the feathers of birds as winter approaches?*

Because the down is a non-conductor of heat, and black the warmest colour. It is therefore best adapted to *keep in* their bodily warmth during the cold of winter.

1037. *Why has man no external appendage to his mouth?*

Because *his hands* serve all the purposes of gathering food, and *conveying it to the mouth.* Man's mouth is simply an *opening;* in other animals it is a *projection.*

1038. *Why have dogs, and other carnivorous animals, long pointed teeth, projecting above the rest?*

Because as they have not hands to seize and controul their food, the projecting teeth enable them to *snap and hold* the objects which they pursue for food.

1039. *Why is the under jaw of the hog, shorter and smaller than the upper one?*

Because the animal pierces the ground *with its long snout,* and then the small under jaw *works freely in the furrow* that has been opened, in quest of food.

1040. *Why have birds hard beaks?*

Because, having no teeth, the beak enables them to *seize, hold,* and *divide their food.*

1041. *Why are the beaks of birds generally long and sharp?*

Because the greater number of birds live by *picking up small*

"As the fishes that are taken in an evil net, and as the birds that are caught in the snare; so are the sons of men snared in an evil time, when it falleth suddenly upon them."—ECCLESIASTES IX.

objects, such as worms, insects, seeds, &c. The sharp beak, therefore, serves as a *fine pincers*, enabling them to take hold of their food conveniently.

1042. *Why have snipes and woodcocks long tapering bills?*

Because they live upon worms which they find in the soft mud of streams and marshy places; their long bills, therefore, enable them to *dig down into the mud after their prey.*

1043. *Why have woodcocks, snipes, &c., nerves running down to the extremities of their bills?*

Because, as they dig for their prey in the soft sand and mud, they cannot see the worms upon which they live. Nerves are, therefore, distributed to the very point of their bills (where, in other birds, nerves are entirely absent) *to enable them to prehend their food.*

Fig. 67.—SPOONBILL.

1044. *Why have ducks and geese square-pointed bills?*

Because they not only feed by dabbling in soft and muddy soil, but they consume a considerable quantity of green food, and their square bills enable them to *crop off the blades of grass.*

"Let the heaven and earth praise him, the seas, and everything that moveth therein."—PSALM LXIX.

1045. *Why has the spoon-bill a long expanded bill, lined internally with sharp muscular points?*

Because the bird *lives by suction*, dipping its broad bill in search of acquatic worms, mollusks, insects and the roots of weeds. The bill forms *a natural spoon*, and the muscular points enable the bird to *filter the mud*, and to retain the nourishment which it finds.

1046. *Why has the spoon-bill long legs?*

Because it *wades in marshy places* to find its food. Its legs are therefore long, for the purpose of keeping its body out of the water, and above the smaller acquatic plants, while it searches for its prey.

1047. *Why have the parrots, &c., crooked and hard bills?*

Because they live upon nuts, the stones of fruit, and hard seeds. The shape of the bill, therefore, enables them to *hold the nut or seed firmly*, and the sharp point enables them to *split or remove the husks*.

1048. *Why can a parrot move its upper as well as its lower bill?*

Because by that means it is enabled to bring the nut or seed nearer the fulcrum, or joint of the jaw. It, therefore, acquires greater power, just as with a pair of nut-crackers we obtain increased power by *setting the nut near to the joint*.

1049. *Why have animals with long necks large throats?*

Animals that graze, or feed from the ground, generally have a more powerful muscular formation of the throat than those which feed in other positions, because a greater effort is required to *force the food upward, that would be needed to convey it down*.

1050. *Why are the bones of birds hollow?*

Because they are thereby rendered *lighter*, and do not interfere with the flight of the bird *as they would do if they were solid*. Greater strength is also obtained by the *cylindrical form of the bone*, and a larger surface afforded for the *attachment of powerful muscles*.

"And my hand hath found, as a nest, the riches of the people; and as one gathereth eggs that are left, have I gathered all the earth; and there was none that moved the wing, or opened the mouth, or peeped."—ISAIAH X.

1051. *Why do all birds lay eggs?*

Because, to bear their young in any other manner, would *encumber the body*, and materially interfere with their powers of flight.

As soon as an egg becomes large and heavy enough to be cumbersome to the bird, it is removed from the body. A shell, impervious to air, protects the germ of life within, until from two to twenty eggs have accumulated, and then, although laid at different intervals, their incubation commences together, and the young birds are hatched at the same time.

CHAPTER LIII.

1052. *Why have birds with long legs short tails?*

Because the tails of birds are used to guide them through the air, by *a kind of steerage.* When birds with long legs take to flight, they throw their legs behind, and they then *serve the same purpose as a tail.*

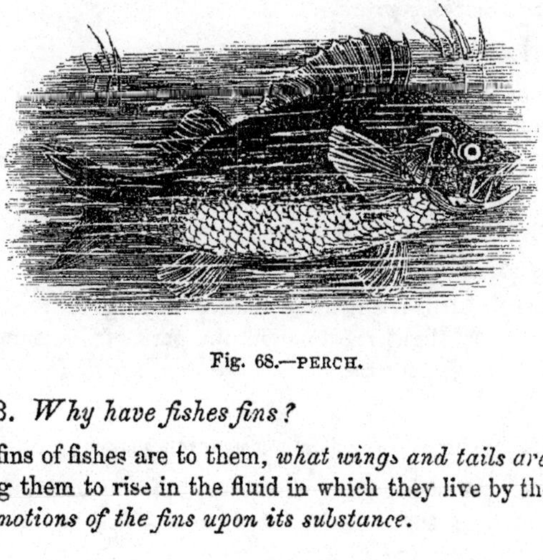

Fig. 68.—PERCH.

1053. *Why have fishes fins?*

The fins of fishes are to them, *what wings and tails are to birds,* enabling them to rise in the fluid in which they live by the *reaction of the motions of the fins upon its substance.*

"Speak to the earth, and it shall teach thee; and the fishes of the sea shall declare unto thee. Who knoweth not in all these that the hand of the Lord hath wrought this."—JOB XII.

1054. *Why are the fins of fishes proportionately so much smaller than the wings of birds?*

Because there is less difference between the *specific gravity* of the body of a fish, and the water in which it moves, than between the body of a bird, and the air on which it flies. The fish, therefore *does not require such an expanded surface to elevate or guide it.*

1055. *Why have fishes scales?*

Because scales, while they afford protection to the bodies of fish, are conveniently adapted to their motions; and as the scales *present no surface to obstruct their passage through the water,* as hair or feathers would do, they evidently form the best covering for the acquatic animal.

1056. *Why do fishes float in streams (when they are not swimming) with their heads towards the stream?*

Because they *breathe* by the transmission of water over the surface of their gills, the water entering at the mouth, and passing over the gills behind. When, therefore, they lie motionless with their heads to the stream, they are in *that position which naturally assists their breathing process.*

1057. *Why have fishes air-bladders?*

Because, as the density of water varies greatly at different depths, the enlargement or contraction of the bladder regulates the relation of *the specific gravity of the body of the fish to that of the water in which it moves.*

1058. *Why have whales a very large development of oily matter about their heads?*

Because their heads are thereby rendered the lighter part of their bodies, and a very slight exertion on the part of the animal will bring its head to the surface *to breathe air, which it constantly requires.*

1059. *Why have birds that swim upon water web-feet?*

Because the spreading out of the toes of the bird brings the membrane between the toes into the form of a fin, or *water-wing,*

"And Jesus saith unto him, The foxes have holes, and the birds of the air have nests; but the son of man hath not where to lay his head."—MATTHEW XIII.

by striking which against the water, *the bird propels itself along.*

1060. *Why have birds that swim and dive short legs?*

Because long legs would greatly *impede their motions in the water*, by becoming repeatedly entangled in the weeds, and by striking against the bottom. *Waders*, however, require long legs because they have to move about through the *tall vegetation of marshy borders.*

Fig. 69—STILT-PLOVER AND DUCK.

1061. *Why have the feet of the heron, cormorant, &c., deep rough notches upon their under surface?*

Because, as those birds live by catching fish, they are enabled by the notches in their feet, to *hold the slippery creatures upon which they feed.*

1062. *Why have otters, seals, &c., web-feet?*

Because, while the feet enable them to *walk upon the land*, they are equally effective in their action upon the water, and hence they are *adapted to the amphibious nature of the animals to which they belong.*

1063. *Why do the external ears of animals of prey, such as cats, tigers, foxes, wolves, hyenas, &c., bend forward?*

Because they collect the sounds that occur *in the direction of the*

"Doth the hawk fly by thy wisdom, and stretch her wings toward the south?
"Doth the eagle mount up at thy command, and make her nest on high?

pursuit, and enable the animal to *track its prey* with greater certainty.

1064. *Why do the ears of animals of flight, such as hares, rabbits, deer, &c., turn backward?*

Because they thereby catch the sounds that give them *warning of the approach of danger.*

1065. *Why has the stomach of the camel a number of distinct bags, like so many separate stomachs?*

Because water is stored up in the separate chambers of the stomach, apart from the solid aliment, so that the animal can *feed*, without consuming all its drink. It is thereby *able to retain water to satisfy its thirst while travelling across hot deserts*, where no water could be obtained.

1066. *Why do woodpeckers "tap" at old trees?*

Because by boring through the decayed wood, with the sharp and hard bills with which they are provided, *they get at the haunts of the insects upon which they feed.*

1067. *Why are woodpeckers' tongues about three times longer than their bills?*

Because, if their bills were long, they would not bore the trees so efficiently; and when the trees are bored, and the insects alarmed, they endeavour to retreat into the hollows of the wood; *but the long thin tongue of the woodpecker fixes them on its sharp horny point*, and draws them into the mouth of the bird.

1068. *Why have the Indian hogs large horns growing from their nostrils and turning back towards their eyes?*

Because the horns *serve as a defence to the eyes* while the animal forces its way through the thick underwood in which it lives.

1069. *Why have calves and lambs, and the young of horned cattle generally, no horns while they are young?*

Because the presence of horns would *interfere with the suckling*

"She dwelleth and abideth on the rock, upon the crag of the rock, and the strong place.
"From thence she seeketh the prey, and her eyes behold afar off. Her young ones also suck up blood: and where the slain are, there is she."—JOB XXXIX.

of the young animal. When, however, it is able to feed itself by browsing, *then the horns begin to grow.*

1070. *Why have infants no teeth?*

Because the presence of teeth would interfere with their suckling, while the teeth would be of no service, until the child *could take food requiring mastication.*

1071. *Why cannot flesh-eating animals live upon vegetables?*

Because the gastric juice of a flesh-eating animal, being adapted to the duty which it has to perform, *will not dissolve vegetable matter.*

1072. *Why have birds gizzards?*

Because, having no teeth, the tough and fibrous gizzards are employed *to grind the food preparatory to digestion.*

1073. *Why are small particles of sand, stone, &c., found in the gizzards of birds?*

Because, by the presence of those rough particles, which become embedded in the substance of the gizzard, the food of the bird is more effectively ground.

When our fowls are abundantly supplied with meat, they soon fill their craw, but it does not immediately pass thence into the gizzard; it always enters in small quantities, in proportion to the progress of trituration, in like manner, as in a mill, a receiver is fixed above the two large stones which serve for grinding the corn, which receiver, although the corn be put into it by bushels, allows the grain to dribble only in small quantities into the central hole in the upper mill-stone.—*Paley.*

CHAPTER LIV.

1074. *Why has the mole hard and flat feet, armed with sharp nails?*

Because the animal is thereby enabled to *burrow in the earth,* in search for worms. Its feet are so many *shovels.*

1075. *Why is the mole's fur exceedingly glossy and smooth?*

Because its smoothness enables it to work under ground *without the soil sticking to its coat,* by which its progress would be impeded. From soils of all kinds, the little worker emerges shining and clean.

What I have always most admired in the mole is its *eyes*. This animal occasionally visiting the surface, and wanting, for its safety and direction, to be informed when it does so, or when it approaches it, a perception of light was necessary. I do not know that the clearness of sight depends at all upon the size of the organ. What is gained by the largeness or prominence of the globe of the eye, is width in the field of vision. Such a capacity would be of no use to an animal which was to seek its food in the dark. The mole did not want to look about it; nor would a large advanced eye have been easily defended from the annoyance to which the life of the animal must constantly expose it. How indeed was the mole, working its way under ground, to guard its eyes at all? In order to meet this difficulty, the eyes are made scarcely larger than the head of a corking-pin; and these minute globules are sunk so deeply in the skull, and lie so sheltered within the velvet of its covering, as that any contraction of what may be called the eye-brows, not only closes up the apertures which lead to the eyes, but presents a cushion, as it were, to any sharp or protruding substance which might push against them. This aperture, even in its ordinary state, is like a pin-hole in a piece of velvet, scarcely pervious to loose particles of earth. —*Paley.*

FIG. 70.—ELEPHANTS DRINKING.

1076. *Why has the elephant a short unbending neck?*

Because the elephant's head is so heavy, that it could not have been supported at the end of a long neck (or lever), without *a provision of immense muscular power.*

"Be not afraid, ye beasts of the field: for the pastures of the wilderness do spring, for the tree beareth her fruit, the fig-tree and the vine do yield their strength."—JOEL II.

1077. *Why has the elephant a trunk?*

The trunk of an elephant *serves as a substitute for a neck*, enabling the animal to crop the branches of trees, or to raise water from the stream.

1078. *Why do the hind legs of elephants bend forward?*

Because the weight of the animal is so great, that when it lay down it would *rise with great difficulty*, if its legs bent outward, as do the legs of other animals. Being bent *under the body*, they have a greater power of pushing directly upward, when the powerful muscles of the thighs straighten them.

According to Cuvier, the number of muscles, in an elephant's trunk, amounts to *forty thousand*, all of which are under the will, and it is to these that the proboscis of this animal owes its flexibility. It can be protruded or contracted at pleasure, raised up or turned to either side, coiled round on itself or twined around any object. With this instrument the elephant collects the herbage on which he feeds and puts it into his mouth; with this he strips the trees of their branches, or grasps his enemy and dashes him to the ground. But this admirable organ is not only adapted for seizing or holding substances of magnitude; it is also capable of plucking a single leaf, or of picking up a straw from the floor. The orifices of the canals of the extremity are encircled by a projecting margin, produced anteriorly into a finger-like process endowed with a high degree of sensibility and exceedingly flexible. It is at once a finger for grasping and a feeler: the division between the two nasal orifices or their elevated sides serves as a point against which to press; and thus it can pick up or hold a small coin, a bit of biscuit, or any trifling thing with the greatest ease.—*Knight's Animal Kingdom.*

1079. *Why have bats hooked claws in their wings?*

Because bats are almost destitute of legs and feet; at least those organs are included in their wings. If they alight upon the ground, they have great difficulty in again taking to the wing, as they cannot run or spring to bring their wings in action upon the air. At the angle of each wing there is placed, therefore, a bony hook, by which the bat attaches itself to the sides of rocks, caves, and buildings, laying hold of crevices, joinings, chinks, &c.; and when it takes its flight, *it unhooks itself, and its wings are at once free to strike the air.*

1080. *Why does the bat fly by night?*

Because it lives chiefly upon moths, which are *night-flying insects.*

"So are the paths of all that forget God; and the hypocrite's hope shall perish: Whose hope shall be cut off, and whose trust shall be a spider's web."—JOB VIII.

1081. *Why does the bat sleep during the winter?*

Because, as the winter approaches, the moths and flying insects upon which it feeds, disappear. *If, therefore, it did not sleep through the winter, it must have starved.*

Fig. 71.—BAT WITH HOOKED WINGS.

1082. *Why has the spider the power of spinning a web?*

Because, as it lives upon flies, but is *deficient of the power of flying in pursuit of them,* it has been endowed with an instinct *to spread a snare to entrap them,* and with the most wonderful machinery to give that instinct effect.

There are few things better suited to remove the disgust into which young people are betrayed on the view of some natural objects, than this of the spider. They will find that the most despised creature may become a subject of admiration, and be selected by the naturalist to exhibit the marvellous works of the creation. The terms given to these insects, lead us to expect interesting particulars concerning them, since they have been divided into vagrants, hunters, swimmers, and water spiders, sedentary, and mason-spiders, thus evincing a variety in their condition, activity, and mode of life; and we cannot be surprised to find them varying in the performance of their vital functions (as, for example, in their mode of breathing), as well as in their extremities and instruments. Of these instruments the most striking is the apparatus for spinning and weaving, by which they not only fabricate webs to entangle their prey, but form cells for their residence and concealment; sometimes living in the ground, sometimes under water, yet breathing the atmosphere. Corresponding with their very singular organisation are their instincts. We are familiar with the watchfulness and voracity of some spiders, when their prey is indicated by the vibration of the cords of their net-work. Others have the eye and disposition of the lynx or tiger, and after couching in concealment, leap upon their victims. Some conceal themselves under a silken hood or tube, six eyes only projecting. Some bore a hole in the earth, and line it as finely as if it were done with the trowel and mortar, and then hang it with delicate curtains. A very extraordinary degree of contrivance is exhibited in the trap-door spider. This door, from which it derives its name, has a frame and hinge on the mouth of the cell, and is so provided that the claw of the spider can lay hold of it, and whether she enters or goes out, says Mr. Kirby, the door shuts of itself. But the water-spider has a domicile more curious still: it is under water with an opening at the lower part for her exit and entrance, and although this cell be under water, it contains air like a diving-bell, so that the spider breathes the atmosphere. The air is renewed in the cell in a manner not easily explained. The spider comes to the surface; a bubble of air is attracted to its body; with this air she descends, and gets under her cell, when the air is disengaged and rises into the cell; and thus, though under water, she lives in the air. There must be some peculiar property of the surface of this creature by which she can move in the water surrounded with an atmosphere, and live under the water breathing the air.

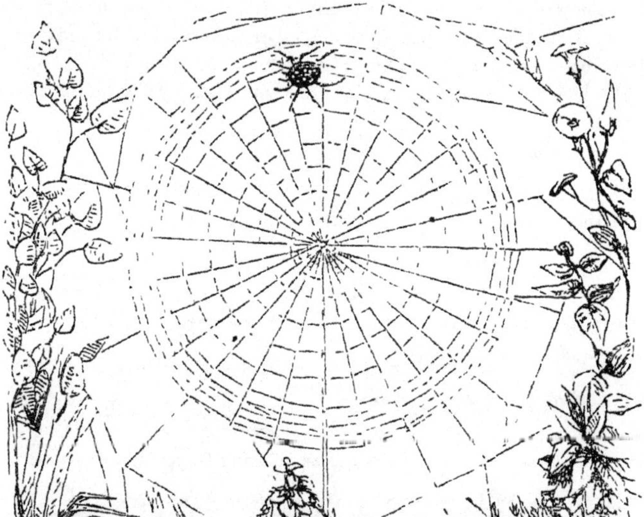

FIG. 72.—WEB OF THE GEOMETRICAL SPIDER.

The chief instrument by which the spider performs these wonders is the spinning apparatus. The matter from which the threads are spun is the liquid contained in cells; the ducts from these cells open upon little projecting teats, and the atmosphere has so immediate an effect upon this liquid, that upon exposure to it the secretion becomes a tough and strong thread. Twenty-four of these fine strands form together a thread of the thickness of that of the silk-worm. We are assured that there are three different sorts of material thus produced, which are indeed required for the various purposes to which the

"For every beast of the forest is mine, and the cattle upon a thousand hills."—
PSALM L.

are applied—as, for example, to mix up with the earth to form the cells; to line these cells as with fine cotton; to make light and floating threads by which they may be conveyed through the air, as well as those meshes which are so geometrically and correctly formed to entrap their prey.—*Note by Lord Brougham to Paley's Natural Theology.*

1083. *Why have many insects a great number of eyes?*

Because the orb of the eye is fixed; there is therefore placed over the eye a multiple-lens, which conducts light to the eye from every direction; so that *the insect can see with a fixed eye as readily as it could have done with a moveable one.* As many as fourteen hundred eyes, or inlets of light, have been counted in the head of a drone-bee. The spider has *eight eyes,* mounted upon different parts of the head; two in front, two in the top of the head, and two on each side.

1084. *Why have birds of prey no gizzards?*

Because their food *does not require to be ground* prior to digestion, as does the food of grain-eating birds.

1085. *Why have earth worms no feet?*

Because the undulatory motion of their muscles serves them for all the purposes of progression *needed by their mode of life.*

1086. *Why have mussels strong tendinous threads proceeding from their shells?*

Because as they live in places that are beaten by the surf of the sea, they *moor their shells* by those threads to rocks and timbers.

1087. *Why have cockles stiff muscular tongues?*

Because, having no threads to moor themselves, as the mussels have, they *dig out with their tongues a shelter for themselves in the sand.*

1088. *Why do oxen, sheep, deer, &c., ruminate?*

Because they have no front teeth in the upper jaw, the place of which is occupied by a hardened gum. The first process, therefore, consists simply of *cropping* their food, which is passed into the paunch, to *be brought up again and ground by the back teeth* when the cropping process is over.

Because, in a wild state, they are constantly exposed to the

"A righteous man regardeth the life of his beast: but the tender mercies of the wicked are cruel."—PROVERBS XII.

attacks of carnivorous beasts, and as the mastication of the large amount of vegetable food required for their sustenance would take a considerable time, they are provided with stomachs, by which they are enabled to fill their paunches quickly, and then, retiring to a place of safety, they bring their food up again, and chew it at leisure.

1089. *Why can ruminating animals recover the food from their paunches?*

Because they have a *voluntary power* over the muscles of the throat, by which they can bring up the food at will.

1090. *Why can they keep the unchewed food in the paunch, from the "cud" they have chewed for nourishment?*

Because their stomachs are divided into three chambers: 1, the *paunch*, where the unchewed food is stored; 2, the *reticulum*, where portions of the food are received from the paunch, and moistened and rolled into a "cud," to be sent up and chewed; and 3, the *psalterium*, which receives the masticated food, and continues the process of digestion.

In quadrupeds the deficiency of teeth is usually *compensated* by the faculty of rumination. The sheep, deer, and ox tribe, are without fore-teeth in the upper jaw. These ruminate. The horse and ass are furnished with teeth in the upper jaw, and do not ruminate. In the former class, the grass and hay descend into the stomachs nearly in the state in which they are cropped from the pasture, or gathered from the bundle. In the stomach, they are softened by the gastric juice, which in these animals is unusually copious. Thus softened and rendered tender, they are returned a second time to the action of the mouth, where the grinding teeth complete at their leisure the trituration which is necessary; but which was before left imperfect. I say, the trituration which is necessary; for it appears from experiments, that the gastric fluid of sheep, for example, has no effect in digesting plants, unless they have been previously masticated; that it only produces a slight maceration, nearly as common water would do in a like degree of heat; but that when once vegetables are reduced to pieces by mastication, the fluid then exerts upon them its specific operation. Its first effect is to soften them, and to destroy their natural consistency; it then goes on to dissolve them, not sparing even the toughest parts, such as the nerves of the leaves. I think it very probable, that the gratification also of the animal is renewed and prolonged by this faculty. Sheep, deer, and oxen, appear to be in a state of enjoyment whilst they are chewing the cud. It is then, perhaps, that they best relish their food.—*Paley*.

"I am like a pelican of the wilderness: I am like an owl of the desert. I watch, and am as a sparrow alone upon the house top."—PSALM CII.

CHAPTER LV.

1091. *Why do quadrupeds that are vegetable eaters feed so continually?*

Because their food contains but a *small proportion of nutrition*, so that it is necessary to digest a *large quantity* to obtain sufficient nourishment.

1092. *Why do flesh eating animals satisfy themselves with a rapid meal?*

Because the food which they eat is *rich in nutritious matter*, and more readily digestible than vegetable food; it does not therefore, require the same amount of *grinding with the teeth.*

Fig. 73.—PELICAN WITH DILATED POUCH.

1093. *Why has the pelican a large pouch under its bill?*

Because it subsists upon fish, generally of the smaller kind, and uses its pouch *as a net* for catching them; the pouch also serves as a *paunch,* in which the fish are stored, until the bird ceases from the exertion of fishing, and takes its meal at leisure.

In their wild state they hover and wheel over the surface of the water, watching the shoals of fish beneath, and suddenly sweeping down, bury themselves in the foaming waves; rising immediately from the water by their own buoyancy, up they soar, the pouch laden with the fish scooped up during their momentary submersion. The number of fish the pouch of this species will contain may be easily imagined when we state that it is so dilatable as to be capable of containing two gallons of water, yet the bird has the power of contracting this membranous expansion, by wrinkling it up under the lower mandible, until it is scarcely to be seen. In shallow inlets, which the pelicans often frequent, it nets its prey with great adroitness.

The pelican chooses remote and solitary islands, isolated rocks in the sea, the borders of lakes and rivers, as its breeding place. The nest, placed on the ground, is made of coarse grasses, and the eggs, which are white, are two or three in number. While the female is incubating, the male brings fish to her in his pouch, and the young, when hatched, are assiduously attended by the parents, who feed them by pressing the pouch against the breast, so as to transfer the fish from the former into the throats of the young. This action has doubtless given origin to the old fable of the pelican feeding its young with blood drawn from its own breast.—*Knight's Animal Kingdom.*

1094. *Why do the smaller animals breed more abundantly than the larger ones?*

Because the smaller ones are designed to be the food of the larger ones, and are therefore *created in numbers adapted to that end.* An elephant produces but one calf; the whale but one young one; a butterfly lays six hundred eggs; silk-worms lay from 1,000 to 2,000 eggs; the wasp, 5,000; the ant, 4,000 to 5,000; the queen bee, 5,000 to 6,000, or 40,000 to 50,000 in a season; and a species of white ant *(termes fatalis)* produces 86,400 eggs in a day. Birds of prey seldom produce more than two eggs; the sparrow and duck tribe frequently sit upon a dozen; in rivers there prevail a thousand minnows for one pike; and in the sea, a million of herrings for a single shark; while of the animalcules upon which the whale subsists, there must exist hundreds of millions for one whale.

1095. *Why has the whale feathery-like laminæ of whalebone extending from its jaws?*

Because these feathery bones, lying side by side, form a *sieve, or strainer,* for the large volumes of water which the whale receives into its its mouth, drawing off therefrom millions of small animals,

"Hast thou given the horse strength? hast thou clothed his neck with thunder? * * He paveth the valley, and rejoiceth in his strength: he goeth on to meet the armed men."—JOB XXXIX.

which form a jelly-like mass upon which the whale feeds. A whale has been known to weigh as much as 249 tons, and its blubber yielded 4,000 gallons of oil. How many millions of living creatures must have gone to make up that enormous mass of animal matter!

1096. *Why have cats, and various other animals, whiskers?*

The whiskers of cats, and of the cat tribe, are exceedingly sensitive, enabling them, when seizing their prey in the dark, to *feel its position most acutely*. These hairs are supplied, through their roots, with branches of the same nerves that give sensibility to the lips, and that in insects *supply their "feelers."*

1097. *Why has the horse a smaller stomach proportionately than other animals?*

Because the horse was created for speed. Had he the ruminating stomach of the ox, he would be quite unfitted for the labour which he now so admirably performs.

1098. *Why has the horse no gall-bladder?*

Because the rapid digestion of the horse, by which its fitness for speed is greatly increased, *does not require the storing up of the bile* as in other animals in which the digestive process is a slower operation.

1099. *Why do certain butterflies lay their eggs upon cabbage leaves?*

Because the cabbage leaves are *the food of the young caterpillars;* and although the butterfly does not subsist herself upon the leaf, she knows by instinct that the leaf will afford food to her future young; she therefore lays her eggs where her young ones will find food.

This explanation applies to many insects that lay their eggs upon other plants

1100. *Why have insects long projections from their heads, like horns or feathers?*

Because those organs (the *antennæ*), are those through which some insects *hear* and others *feel;* and the projecting of these *antennæ* from their bodies probably enables them to hear or feel

"My son, eat thou honey, because it is good; and the honey-comb, which is sweet to thy taste."—PROVERBS XXIV.

more acutely while their wings are in motion, *without the interference of the vibrations of their wings.*

1101. *Why have bees stings?*

Because they gather and store up honey which would constantly attract other insects, and the bees would be robbed of their food but for the sting, *which is given to them for protection.*

1102. *Why have flies fine hairs growing at the extremities of their legs?*

Because they require to cleanse their bodies and wings, and to free them from particles of dust. And as they cannot turn their heads for this purpose, they have hairy feet, which serve as brushes, by which any part of their bodies can be reached and cleaned.

CHAPTER LVI.

1103. *Why when the perfume of flowers is unusually perceptible may wet weather be anticipated?*

Because when the air is damp it *conveys the odours of flowers* more effectively than it does when dry.

1104. *Why when swallows fly low may wet weather be expected?*

Because the insects which the swallows pursue in their flight are flying low, *to escape the moisture of the upper regions of the atmosphere.*

1105. *Why do ducks and geese go to the water, and dash it over their backs on the approach of rain?*

Because by wetting the outer coat of their feathers before the rain falls, by sudden dashes of water over the surface, they *prevent the drops of rain from penetrating to their bodies through the open and dry feathers.*

1106. *Why do horses and cattle stretch out their necks and snuff the air on the approach of rain?*

Because they smell the *fragrant perfume* which is diffused in the air by its increasing moistness.

"I will remember the works of the Lord: Surely I will remember thy wonders of old."—PSALM LXXVII.

1107. *Why may change of weather be anticipated when domestic animals are restless?*

Because their skins are exceedingly sensitive to atmospheric influences, and they are oppressed and irritated by *the changing condition of the atmosphere.*

1108. *Why may fine weather be expected when spiders are seen busily constructing their webs?*

Because those insects are highly sensitive to the state of the atmosphere, and when it is setting fine they build their webs, because they know instinctively *that flies will be abroad.*

1109. *Why is wet weather to be expected when spiders hide?*

Because it shows that they are aware that the state of the atmosphere does not *favour the flight of insects.*

1110. *Why if gnats fly in large numbers may fine weather be expected?*

Because it shows that they feel the state of the atmosphere to be favourable, which induces them all to *leave their places of shelter.*

1111. *Why if owls scream during foul weather, will it change to fine?*

Because the birds are pleasurably excited by a favourable *change in the atmosphere.*

1112. *Why is it said that the moping of the owl foretels death?*

Because owls scream when the weather is on the change; and when a patient is lingering on a death bed, the alteration in the state of the atmosphere frequently induces death, because the faint and expiring flame of life has not strength enough *to adapt itself to the change.*

1113. *Why may wet weather be expected when spiders break off their webs, and remove them?*

Because the insects, anticipating the approach of rain, remove their webs for preservation.

"There shall the great owl make her nest, and lay, and hatch, and gather under her shadow: there shall the vultures also be gathered, every one with her mate."—ISAIAH XXXIV.

1114. *Why may we expect a continuance of fine weather when bees wander far from their hives?*

Because the bees feel instinctively that from the state of the atmosphere they may wander far in search of honey, without the danger of being overtaken by rain.

1115. *Why if people feel their corns ache, and their bones rheumatic, may rain be expected?*

Because the dampness of the atmosphere affects its pressure upon the body, and causes a temporary disturbance of the system. All general disturbances of the body, *manifest themselves in those parts which are in a morbid state*—as in a corn, a rheumatic bone, or a decayed tooth.

1116. *Why if various flowers close may rain be expected?*

Because plants are highly sensitive to atmospheric changes, and *close their petals to protect their stamens.*

1117. *Why when moles throw up their hills may rain be expected?*

Because the moles know instinctively, that on the approach of wet, worms move in the ground; the moles therefore become active, *and form their hills.*

1118. *Why is a magpie, when seen alone, said to fortell bad weather?*

Because magpies generally fly in company; but on the approach of wet or cold, one *remains in the nest to take care of the young,* while the other one wanders alone in search of food.

1119. *Why do sea-gulls appear numerous in fine weather?*

Because the fishes swim near to the surface of the sea, and the birds *assemble over the sea to catch the fish, instead of sitting on rocks,* or wading on the shore.

1120. *Why do sea-gulls fly over the land, on the approach of stormy weather?*

Because in stormy weather they cannot catch fish; and the *earth-worms come up on the land* when the rain falls.

"And I said, Oh, that I had wings like a dove! for then would I fly away, and be at rest."—PSALM LV.

1121. *Why if birds cease to sing, may wet, and probably thunder, be expected.*

Because birds are depressed by an unfavourable change in the atmosphere, and *lose those joyful spirits which give rise to their songs.*

1122. *Why if cattle run around in meadows, may thunder be expected?*

Because the electrical state of the atmosphere has the effect of making them feel uneasy and irritable, and *they chase each other about to get rid of the irritability.*

1123. *Why if birds of passage arrive early, may severe weather be expected?*

Because it shows that the indications of unfavourable weather have set in, in the latitudes from which the birds come, and that they have *taken an early flight to escape it.*

1124. *Why if the webs of the gossamer spider fly about in the autumn, may east winds be anticipated?*

Because an east wind is a dry and dense wind, and suitable to the flight of the gossamer spider; the spider feeling instinctively the dryness of the air, throws out its web, and finds it *more than usually buoyant upon the dense air.*

The observation of the changing phenomena which attend the various states of the weather is a very interesting study, though no general rules can be laid down that can be relied upon, because there are modifying circumstances which influence the weather in various localities and climates. To observe weather indications accurately, no phenomenon should be taken alone, but several should be regarded together. The character and the duration of the weather of the preceding days, the direction of the wind, the forms of the clouds, the indications of the barometer, the rise or fall of the thermometer, and the instinctive forewarnings of birds, beasts, insects, and flowers, should all be taken into account. Although no direct material advantages attend such a study, it induces a habit of observation, and developes the inductive faculty of the mind, which, when applied to more significant things, may trace important effects to their greater causes.

"Go to the ant, thou sluggard; consider her ways, and be wise."—PROV. VI.

CHAPTER LVII.

1125. *Why can gossamer spiders float through the air?*

Because, having no wings, and being deficient in the active muscular powers of other spiders, they have been endowed with the power of spinning a web which is so light that it floats in the air, and bears the body of the gossamer spider from place to place. Each web acts as a balloon, and the spider attached thereto is a little *aeronaut.*

1126. *Why do crickets make a peculiar chirping sound?*

Because they have hard wing cases, by the friction of the edges of which they cause their peculiar noise, to *make known to each other where they are,* in the dark crevices in which they hide.

FIG. 74.—GLOW-WORM USING HIS BRUSH.

1127. *Why has the glow-worm a brush attached to its tail?*

Because it is necessary to keep its back very clean, that *the light which its body emits may not be dimmed.*

1128. *Why does the glow-worm emit a light?*

Because the female glow-worm is without wings, but the male is a winged insect. The female, therefore, is endowed with the power of displaying a phosphorescent light. The light is only visible by night, but it is, nevertheless, beautifully adapted for the purpose stated, because *the male is a night-flying insect,* and never ventures abroad by day.

There exists some difference of opinion between naturalists upon the uses of the light of a glow-worm; there are some who doubt that it is exhibited to attract the flying insect. The objectors, however, offer no explanation of the luminous properties of the worm. Sir Charles Bell says the preponderance of the argument is decidedly in favour of the explanation we have given.

1129. *Why does not the iris of the fish's eye contract?*

Because the diminished light in water is *never too strong for the retina.*

1130. *Why is the eye of the eel covered with a transparent horny covering?*

Because, as the eel lives in holes, and pushes its head into mud, and under stones, &c., it needed such a covering to *defend the eye.*

1131. *Why is the whale provided with an eye, having remarkably thick and strong coats?*

Because, when he is attacked by the sword-fish and the shark, he is almost helpless against his enemies, as they fix themselves upon his huge carcase. He therefore dives with them down to a depth where the pressure of the water is so great that they cannot bear it. The eye of the whale is expressly organised *to bear the immense pressure of extreme ocean depths,* without impairing the sight.

1132. *Why have fishes no eye-lids?*

Because the water in which they swim keeps their eyes moist. Eyelids would therefore be *useless to them.*

1133. *Why have fishes the power of giving their eye-balls very sudden motion?*

Because, having no eyelids (such organs being unnecessary to keep their eyes moist), they still need the power of freeing their eyes from the contact of foreign matters; and this is secured to them by the power they have of giving the eye-ball a very rapid motion, which causes reaction in the fluid surrounding it, and *sweeps the surface.*

This motion may frequently be seen in the eyes of fishes, in glass globes.

"And God made the beast of the earth after his kind, and cattle after their kind, and everything that creepeth upon the earth after his kind: and God saw that it was good."—GENESIS I.

1134. *Why is the lachrymal secretion of the horse's eye thick and glutinous?*

Because, as his eye is large, and constantly exposed to dust on journies, it is provided with a *viscid secretion,* which cleanses the eye, and more instantly and securely removes the dust, than a *watery* secretion would.

1135. *Why does the lower bill of the sea-crow project beyond the upper one?*

Because the bird obtains his food by *skimming along the water,* into which he dips his bill, and lifts his food out.

1136. *Why do the mandibles of the cross-bill overlap each other?*

Because the bird requires a peculiar bill, to enable it to *split seeds into halves,* and to tear the open cones of the fir-tree.

1137. *Why are the tails of fishes so much larger than their fins?*

Because their tails are their *chief instruments of motion,* while their fins are employed simply to direct their progress, and steady their movements.

1138. *Why have oxen, and other quadrupeds a tough ligament called the "pax-wax," running from their backs to their heads?*

Because their heads are of considerable weight; and having frequent occasion to lift them, they are provided with an elastic ligament, which is fastened at the middle of their backs, while its other extremity is attached to the head. This enables them to raise their heads easily; otherwise the effort to do so would be a work of great labour. To the horse, the pax-wax acts as a natural bearing-rein, assisting it to hold its head in that position which adds to the grace and beauty of the animal.

In carving beef, this ligament may be seen passing along the vertebræ of the neck, the chuck, and the fore ribs.

"He shall feed his flock like a shepherd; he shall gather the lambs with his arm, and carry them in his bosom, and shall gently lead those that are with young."—ISAIAH XL.

1139. *Why have the females of the kangaroo and opossum tribes pouches, or pockets, formed in the skin of their breasts for the reception of their young?*

Because their young ones are remarkably *small and helpless;* in fact, more so than those of any other animal of equal proportions. Besides which, the full grown animals have very long hind-legs, and they progress by a series of extraordinary leaps. It would consequently be impossible for their helpless young ones to follow them: God has therefore given to female kangaroos and opossums curious pockets, *formed out of their own skin,* in which they place their little young ones, and bear them through their surprising leaps with the greatest ease and safety.

CHAPTER LVIII.

1140. *What is the difference between an animal, a plant, and a mineral?*

The great naturalist, Linnæus, used to say that animals *grow, live,* and *feel;* plants *grow* and *live;* and minerals *grow.*

Animals are here defined to enjoy *three* conditions of existence; plants *two* conditions; and minerals *one* condition.

This definition has, in latter days, been held to be unsatisfactory, since there *are a few plants* that are *supposed to feel,* and *a few animals* that are supposed to have even *less feeling* than the *sensitive plants* alluded to.

The concise definition by Linnæus, nevertheless, is true, as far as regards a *vast majority* of the bodies constituting the three great kingdoms of nature. And it may be sufficient to say that

Animals—grow, live, feel, and move.

Plants—grow and live.

Minerals—grow, by the addition of particles of inorganic matter.

If we now state the few exceptions that are admitted to this definition, we shall bring the explanation as near to the truth, as the present state of knowledge will permit.

"And God said, Behold, I have given you every herb bearing seed, which is upon the face of all the earth, and every tree, in the which is the fruit of a tree, yielding seed; to you it shall be for meat."—GENESIS I.

1141. *Why is it understood that some plants feel?*

Because the *sensitive plant* closes its leaves on being touched; the *Venus's fly trap* closes its leaves upon flies that alight upon them; others *close* upon the approach of rain, and at sunset, and *open* at sunrise, and turn towards the sun during its daily transit.

1142. *Why is it understood that some plants move?*

Because certain *sea-weeds* throw off undeveloped young plants, which move through the water by the aid of fine *cilia*, or muscular hairs, until they find a suitable place upon which to attach themselves.

The roots of plants will penetrate through the ground in the direction of water, and of favourable soil.

1143. *Of what elementary substances are plants composed?*

Of carbon, oxygen, hydrogen, and nitrogen.

1144. *Whence do plants derive those substances?*

From the air, the earth, and water.

1145. *How do plants obtain carbon?*

They obtain it chiefly from the air, in the form of *carbonic acid gas.* The carbon, of the carbonic acid gas, which is thrown out by the breath of animals, and by other processes in nature, is *absorbed by the leaves of plants*, and the *oxygen* which had united with the carbon to form the *carbonic acid gas*, is again set free for the use of animals.

1146. *How do plants obtain oxygen?*

They obtain it from the *atmospheric air.* But as they do not require a large amount of oxygen for their own use, *they throw off the amount which is in excess*, after having separated it from the other elements with which it was combined when taken up by them. From the humble blade of grass, to the stately tree of the forest, plants operate to purify the air, and to correct and counteract the corruption of the air, by the myriads of animals inhabiting the earth.

It has been generally stated that plants in rooms purify the air by absorbing carbonic acid *by day*, and releasing a part of the oxygen; but that, as the presence of light is necessary to produce this action, they do not restore oxygen to the air, by night, but, on the contrary, give off carbonic acid gas. Therefore it has been stated that plants in rooms by night are unhealthy. Mr. Robert Hunt, one of the ablest chemists of the present time, makes the following remarks upon this subject in his "Poetry of Science:"—

"The power of decomposing carbonic acid is a vital function which belongs to the leaves and bark. It has been stated, on the authority of Leibig, that during the night the plant acts only as a mere bundle of fibres—that it allows of the circulation of carbonic acid and its evaporation, unchanged. In his eagerness to support his chemical hypothesis of respiration, the able chemist neglected to enquire if this was absolutely correct. The healthy plant never ceases to decompose carbonic acid during one moment of its existence; but during the night, when the excitement of light is removed, and the plant reposes, its vital powers are at their minimum of action, and a much less quantity is decomposed than when a stimulating sun, by the action of its rays, is compelling the exertion of every vital function."

In hot, swampy countries, where vegetation is very rapid, and the soil loaded with decomposing carbonic matter, the plants absorb more carbonic acid than they require, and they *then* evolve carbonic acid gas from their leaves. Hence such climates as the West Indies are injurious to *life*, though favourable to *vegetation*.

1147. *How do plants obtain hydrogen?*

They obtain *hydrogen* in combination with *oxygen* in water, and with *nitrogen*, in the form of *ammonia*, as which it exists in animal manures.

1148. *How do plants obtain nitrogen?*

From the *atmospheric air*, and from the *soil*, in which it is combined with other elements.

1149. *How do plants apply these elements to the formation of their own structures?*

When those substances which form the food of plants are absorbed, either by their leaves or their roots, they are converted, with the aid of water, into a *nutritive sap*, which answers the same purposes in *plants* as *blood* does in *animals*.

1150. *How is the nutritive sap applied to the growth and enlargement of the plant?*

Every seed contains a small amount of nutrition, sufficient for the sustentation of the *germ of the plant*, until those vessels are formed, by which the nutritive elements can be absorbed and used for the further development of the living structure.

The earth, penetrated by the sun's rays, warms the sleeping germ, and quickens it into life. For a short time the germ lives upon the seed, which, moistened and warmed by the soil, yields a kind of glutinous sap, out of which the first members of the plant are formed. And then the tender leaf, looking up to the sky, and the slender rootlet penetrating the soil, begin to draw their sustenance from the vast stores of nature.

1151. *Of what do vegetable structures consist?*

Of *membranes*, or thin tissues, which, being variously arranged, form cells, tubes, air passages, &c. Of *fibres*, which form a stronger kind of membrane, and which is variously applied to the production of the organs of the plants. And of *organs*, formed by those elementary substances, by which the plants absorb, secrete, and grow, and fulfil the conditions of their existence.

1152. *Why are seeds generally enveloped in hard cases?*

Because the covering of the seed, like the shell of an egg, is designed *to preserve the germ* within from the influence of external agencies, until the time for development has arrived, and the conditions of germination are fulfilled.

1153. *Why does a seed throw out a root, before it forms a leaf?*

Because moisture, which the root absorbs from the earth, is necessary to enable the germ *to use the nutrition which the seed itself contains*, and out of which the leaf must be eliminated. Moisture forms a kind of gluten, in which the starch of the seed is dissolved, and converted into sugar, the sugar into carbonaceous sap, and the sap into cellular tissue and woody fibre, as the leaves present themselves to the influence of the air and light.

1154. *Why does a plant grow?*

Because, as soon as membranes and vessels are organised in the young germ, the nutritive fluid, formed by its first organs, *begins to move through the fine structures*, and from that time the plant commences to incorporate with its own substance the elements with which it is surrounded, that are suitable to its development.

"Can the rush grow up without mire? can the flag grow without water? Whilst it is yet in his greenness, and not cut down, it withereth before any other herb."—JOB VIII.

CHAPTER LIX.

1155. *Why, if we break the stem of a hyacinth, do we see a glutinous fluid exude?*

Because, by breaking the stem, we rupture the vessels of the plant, and cause the nutritive fluid to escape. The sap of the plant is *analagous to the blood of man,* and the vessels, to the arteries and veins of the animal body.

1156. *Why, if we split the petal of a tulip, do we see cells containing matter of various colours?*

Because, by splitting the petal of the flower, we disclose the anatomy of its structure, and bring to view those cells, or organs, of the vegetable body, by which *the different colouring matters are secreted.*

1157. *Why, if we break a pea-shell across, do we discover a transparent membrane which may be removed from the green cells underneath?*

Because we separate from the cellular, or fleshy part of the shell, the membrane, *which forms the epidermis,* and answers to the skin of the animal body.

1158. *Why, if we cut through a cabbage stump, do we find an outer coat of woody fibre, and an inner substance of cellular matter?*

Because the woody fibre *forms a kind of skeleton,* which supports the internal stricture of the plant, and gives form and character to its organisation. The woody fibre of plants is analagous to the bony structure of animal bodies.

1159. *Why, if we cut across the stem of a plant do we see numerous tubes arranged in parallel lines?*

Because we thereby bring to view *the vessels formed by the membranes and fibres* of the vegetable body, for the transmission of the fluids, by which the structure is sustained.

Skeleton leaves, and seed vessels of plants, form exceedingly interesting

"It was planted in a good soil by great waters, that it might bring forth branches, and that it might bear fruit, that it might be a goodly vine."—EZEKIEL XVII.

objects, and serve to illustrate the wonderful structure of plants. With patience and care, they may be produced by any person, and will afford an interesting occupation. The leaves should be gathered when they are in perfection—that is, when some of the earliest leaves begin to fall from the trees. Select perfect leaves, taking care that they are not broken, or injured by insects. Lay them in pans of *rain water*, and expose them to the air to undergo decomposition. Renew the water from time to time, taking care not to damage the leaves. They need not be examined more than once a week, and then only to see that the water is sufficient to cover them. Give them sufficient time for their soft parts to become decomposed, then take them out, and laying them on a white plate with a little water, wash away carefully, with a camel-hair pencil, the green matter that clings to the fibres. The chief requirement is *patience* on the part of the operator, to allow the leaves and seed vessels sufficient time to decompose. Some leaves will take a few weeks, and others a few months, but a large panful may be put to decompose at the same time, and there will always be some ready for the process of cleansing. When they are thorougly cleaned, they should be bleached, by steeping for a short time in a weak solution of chloride of lime. They should then be dried, and either pressed flat, or arranged in boquets for preservation under glass shades. The result will amply reward the perseverance of the operator.

1160. *Why are clayey soils unfavourable to vegetation?*

Because the soil is *too close and adhesive* to allow of the free passage of air or water to the roots of the plants; it also obstructs the expansion of the fibres of the roots.

1161. *Why are sandy soils unfavourable to vegetation?*

Because they consist of particles that have *too little adhesion to each other*; they do not retain sufficient moisture for the nourishment of the plants; and they allow too much solar heat to pass to the roots.

1162. *Why are chalk soils unfavourable to vegetation?*

Because they do not absorb the solar rays, *and are therefo cold to the roots of plants.*

1163. *Why are mixed soils favourable to vegetation?*

Because they contain the *elements of nutrition* essential to the development of the vegetables, and the plants absorb from them those constituents which are necessary to their growth.

1164. *Why do farmers sow different crops in rotation?*

Because every plant takes something from the soil, and gives

"He watereth the hills from his chambers; the earth is satisfied with the fruit of thy works."—PSALM CIV.

something back; but all kinds of plants do not absorb nor restore the elements in the same proportions. Therefore a succession of crops of one kind would soon impoverish the soil; but a succession of crops of different kinds will compensate the soil, in some degree, for the nourishment withdrawn.

1165. *Why do farmers manure their lands?*

Because, as soils vary, and crops impoverish the soils, the farmer employs manure *to restore fertility*, and to *adapt the soils to the wants of the plants* he desires to cultivate.

It is remarkable that Nature herself points out to man the necessity for changing the succession of vegetable growths.

When plants have exhausted the soil upon which they grow, they will push their roots far in search of sustenance, and in time migrate to a new soil, while other plants will spring up and thrive upon the area vacated. When a forest in North America is destroyed by fire, the trees that grow afterwards are unlike those that the fire consumed, and evidently arise from seeds that have long lain buried in the earth, waiting the time when the ascendancy of the reigning order of plants should cease.

1166. *Why are grasses so widely diffused throughout nature?*

Because they form the *food* of a very large portion of the animal kingdom. They have therefore been abundantly provided. No spot of earth is allowed to remain idle long. When the foot of man ceases to tread down the path, grass immediately begins to appear; and by its universality and the hardihood of its nature, it clothes the earth as with a carpet.

Many grasses, whose leaves are so dry and withered that the plants appear dead, revive and renew their existence in the spring by pushing forth new leaves from the bosom of the former ones.—*Withering's Botany.*

Grasses are Nature's care. With these she clothes the earth; with these she sustains its inhabitants. Cattle feed upon their leaves; birds upon their smaller seeds; men upon the larger; for, few readers need be told that the plants which produce our bread-corn, belong to this class. In those tribes which are more generally considered as grasses, their extraordinary means and powers of preservation and increase, their hardiness, their almost unconquerable disposition to spread, their faculties of reviviscence, coincide with the intention of nature concerning them. They thrive under a treatment by which other plants are destroyed. The more their leaves are consumed, the more their roots increase. The more they are trampled upon, the thicker they grow. Many of the seemingly dry and dead leaves of grasses revive, and renew their verdure in the spring. In lofty mountains, where the summer heats are not sufficient to ripen the seeds, grasses abound which are viviparous, and consequently able to propagate themselves without seed. It is an observation, likewise, which has often been made, that herbivorous animals attach themselves to the leaves of grasses, and, if at liberty in their pastures to range and choose, leave untouched the straws which support the flowers.—*Paley.*

"For the earth bringeth forth fruit of herself; first the blade, then the ear, after that the full ear in the corn."—MARK V.

CHAPTER LX.

1167. *Why do some plants droop, and turn to the earth after sunset?*

Because, when the warmth of the sun's rays is withdrawn, they turn downwards, and *receive the warmth of the earth by radiation.*

1167* *Why does the young ear of corn first appear enfolded in two green leaves?*

Because the light and air would *act too powerfully for the young ear;* two leaves therefore join, and embrace the ear, and protect it until it has acquired strength, when they divide, and leave the ear to swell and ripen.

1168. *Why are the seeds of plants usually formed within the corollas of flowers?*

Because the petals of the flowers, surrounding the seeds, *afford them protection until they are ripened,* when the flower dies, and the petals fall to the ground.

1169. *Why does the flower of the poppy turn down during the early formation of seed?*

Because the heat would probably be too great for the seed in its early stage. The plant is therefore provided with a curious *curve in its stalk,* which turns the flower downward. But when the seeds are prepared for ripening, *the stalk erects itself,* and the seeds are then presented to the ripening influences of the sun.

1170. *Why have plants of the pea tribe, a folding blossom called the "boat," or "keel?"*

Because, within that blossom the pea is formed, and the shape of the blossom is exactly suited to that of the pea which is formed therein. The blossom is itself protected by external petals; and when the wind blows, and threatens to destroy the parts upon which the seeds depend, the plants *turn their backs to the wind,* and shelter the seed.

"The fruit of the righteous is a tree of life; and he that winneth souls is wise."—PROVERBS XI.

1171. *Why are the leaf buds enclosed in scales which fall off as the leaf opens?*

Because the scales *serve as a shelter* to the tender structure of the young leaf. The scales are rudimentary leaves, formed at the end of the previous season, and which, being undeveloped then, serve to guard the young leaves of the future year.

In trees, especially those which are natives of colder climates, this point is taken up earlier. Many of these trees (observe in particular the *ash* and the *horse-chestnut*) produce the embryos of the leaves and flowers in one year, and bring them to perfection the following. There is a winter therefore to be gotten over. Now what we are to remark is, how nature has prepared for the trials and severities of that season. These tender embryos are, in the first place wrapped up with a compactness, which no art can imitate; in which state they compose what we call the bud. This is not all. The bud itself is enclosed in scales; which scales are formed from the remains of past leaves, and the rudiments of future ones. Neither is this the whole. In the coldest climates, a third preservative is added, by the bud having a *coat* of gum or resin, which, being congealed, resists the strongest frosts. On the approach of warm weather this gum is softened, and ceases to be an hinderance to the expansion of the leaves and flowers. All this care is part of that system of provisions which has for its object and consummation, the production and perfecting of the seeds.—*Paley.*

1172. *Why are the seeds of many plants enclosed in a rich juice, or pulp?*

Because the matter by which the seed is surrounded, as well as being intended for the *nourishment and care of the seed*, is designed for the use of man and of animals, by whom the seed is set free to take its place in the earth.

By virtue of this process, so necessary, but so diversified, we have the seed, at length, in stone-fruits and nuts, incased in a strong shell, the shell itself enclosed in a pulp or husk, by which the seed within is, or hath been, fed; or, more generally (as in grapes, oranges, and the numerous kinds of berries), plunged overhead in a glutinous syrup, contained within a skin or bladder; at other times (as in apples and pears) embedded in the heart of a firm fleshy substance; or (as in strawberries) pricked into the surface of a soft pulp.

These and many more varieties exist in what we call *fruits.* In pulse, and grain, and grasses; seeds (as in the pea tribe) regularly disposed in parchment pods, which, though soft and membranous, completely exclude the wet even in the heaviest rains; the pod also, not seldom, (as in the bean), lined with a fine down; at other times (as in the senna) distended like a blown bladder; or we have the seed enveloped in wool (as in the cotton-plant), lodged (as in pines) between the hard and compact scales of a cone, or barricadoed (as in the artichoke and thistle) with spikes and prickles; in mushrooms, placed under a pent-house; in ferns, within slits in the back part of the leaf; or (which is the most general organisation of all) we find them covered by strong, close tunicles, and attached to the stem according to an order appropriated to each plant, as is seen in the several kinds of grains and of grasses.

In which enumeration, what we have first to notice is, unity of purpose under variety of expedients. Nothing can be more *single* than the design; more *diversified* than the means. Pellicles, shells, pulps, pods, husks, skin, scales armed with thorns, are all employed in prosecuting the same intention. Secondly; we may observe, that in all these cases, the purpose is fulfilled within a just and *limited* degree. We can perceive, that if the seeds of plants were more strongly guarded than they are, their greater security would interfere with other uses. Many species of animals would suffer, and many perish, if they could not obtain access to them. The plant would overrun the soil, or the seed be wasted for want of room to sow itself. It is, sometimes, as necessary to destroy particular species of plants, as it is, at other times, to encourage their growth. Here, as in many cases, a balance is to be maintained between opposite uses. The provisions for the preservation of seeds appear to be directed, chiefly against the inconstancy of the elements, or the sweeping destruction of inclement seasons. The depredation of animals, and the injuries of accidental violence, are allowed for in the abundance of the increase. The result is, that out of the many thousand different plants which cover the earth, not a single species, perhaps, has been lost since the creation.

"When nature has perfected her seeds, her next care is to disperse them. The seed cannot answer its purpose, while it remains confined in the capsule. After the seeds therefore are ripened, the pericarpium opens to let them out, and the opening is not like an accidental bursting, but for the most part, is according to a certain rule in each plant. What I have always thought very extraordinary; nuts and shells, which we can hardly crack with our teeth, divide and make way for the little tender sprout which proceeds from the kernel. Handling the nut, I could hardly conceive how the plantule was ever to get out of it. There are cases, it is said, in which the seed-vessel, by en elastic jerk, at the moment of its explosion, casts the seeds to a distance. We all, however, know, that many seeds (those of most composite flowers, as of the thistle, dandelion, &c.) are endowed with what are not improperly called *wings;* that is, downy appendages, by which they are enabled to float in the air, and are carried oftentimes by the wind to great distances from the plant which produces them. It is the swelling also of this downy tuft within the seed-vessel that seems to overcome the resistance of its coats, and to open a passage for the seed to escape.

But the *constitution* of seeds is still more admirable than either their preservation or their dispersion. In the body of the seed of every species of plant, or nearly of every one, provision is made for two grand purposes: first, for the safety of the *germ;* secondly, for the temporary support of the future plant. The sprout, as folded up in the seed, is delicate and brittle beyond any other substance. It cannot be touched without being broken.

Yet in beans, peas, grass-seeds, grain, fruits, it is so fenced on all sides, so shut up and protected, that whilst the seed itself is rudely handled, tossed into sacks, shovelled into heaps, the sacred particle, the miniature plant remains unhurt. It is wonderful, also, how long many kinds of seeds, by the help of their integuments, and perhaps of their oils, stand out against decay. A grain of mustard-seed has been known to lie in the earth for a hundred

"Say not ye, There are four months, and then cometh harvest? behold, I say unto you, Lift up your eyes, and look on the fields; for they are white already to harvest."—JOHN IV.

years; and as soon as it had acquired a favourable situation, to shoot as vigorously as if just gathered from the plant. Then, as to the second point, the temporary support of the future plant, the matter stands thus. In grain, and pulse, and kernels, and pipins, the germ composes a very small part of the seed. The rest consists of a nutritious substance, from which the sprout draws its aliment for some considerable time after it is put forth; viz., until the fibres, shot out from the other end of the seed, are able to imbibe juices from the earth, in a sufficient quantity for its demand. It is owing to this constitution that we see seeds sprout, and the sprouts make a considerable progress, without any earth at all.

From the conformation of fruits alone, one might be led, even without experience, to suppose, that part of this provision was destined for the utilities of animals. As limited to the plant, the provision itself seems to go beyond its object. The flesh of an apple, the pulp of an orange, the meat of a plum, the fatness of the olive, appear to be *more* than sufficient for the nourishing of the seed or kernel. The event shows, that this redundancy, if it be one, ministers to the support and gratification of animal natures; and when we observe a provision to be more than sufficient for one purpose, yet wanted for another purpose, it is not unfair to conclude that both purposes were contemplated together.—*Paley.*

1173. *Why have climbing plants tough curly tendrils?*

Because, *having no woody stalks of their own* to support them, they require to take hold of surrounding objects, and raise themselves from the ground by climbing. Their spiral tendrils are, therefore, so many hands, assisting them to rise from the earth.

1174. *Why does the pea put forth tendrils, and the bean not?*

Because the bean has in its stalk *sufficient woody fibre to support itself*, but the pea has not. We do not know a single tree or shrub having a firm strong stem sufficient for its support which is *also* supplied with tendrils.

1175. *Why do the ears of wheat stand up by day, and turn down by night?*

Because, when the ear is becoming ripe, the cold dew falling into the ear, might *induce blight;* the ears therefore turn down to the earth, and *receive warmth by radiation.*

1176. *Why have grasses, corn, canes, &c., joints, or knots in their stalks.*

Because a long hollow stem would be liable to bend and break. But the joints are so many points where the fibres are bound together, and the structure *greatly strengthened.*

"Then shall the earth yield her increase, and God, even our own God, shall bless us."—PSALM XLVII.

1177. *Why have the berries of the mistletoe a thick viscid juice?*

Because the mistletoe is a *parasitical* plant, growing upon the bark of other trees. It will not grow in the ground; its seeds are therefore filled with an exceedingly sticky substance, which serves to attach them to the bark of trees, to which the berries attach themselves at once, by throwing out tough fibres; and the next year the plant grows.

Fig. 75.—THE MISTLETOE.

1178. *How are the seeds of the mistletoe transferred from its own stem to the bark of trees.*

Various birds, and particularly the *missel thrush,* feed upon the berries. As the bird moves in pursuit of its food, the viscid berries attach themselves to its feathers, and in this way the thrush is the instrument which conveys the seed to the spot to which it adheres, and from which the tree ultimately grows.

1179. *What is the circulation of the sap in plants?*

The circulation of the sap is the movement of the nutritive juices by which the plant is sustained There is a slow uninterrupted movement of the sap from the root through the stems to the leaves, and downwards from the leaves through the bark to the root.

1180. *Why does the sap of plants thus ascend and descend?*

Because it *conveys upward* from the ground some of the matter by which the plant is to be nourished, and which must undergo digestion in the leaves; and it *brings downward* from the leaves the matters absorbed, for the nourishment of the plant, and discharges through the root the substances which the plant cannot use.

The movement of the sap is most active in the spring; but in the depths of the winter it almost ceases.

There are other motions of the sap in plants, which are called *special*, in distinction from the ascending and descending of the sap, which is called *general*, or common to all plants. The special movements of the sap are peculiar to certain plants, in some of which a fluid, full of little green cells, is found to have a rotatory motion; in other plants, a milky fluid is found to move through particular tissues of the vegetable structure.

1181. *Why are the leaves of plants green?*

Because they secrete a carbonaceous matter, named *chlorophyl*, from which they derive their green colour.

1182. *Why are the hearts of cabbages, lettuces, &c., of a pale yellow colour?*

Because the action of *light* is necessary to the formation of *chlorophyl;* and as the leaves are folded upon each other, they exclude the light, and the green matter is not formed.

1183. *Why do leaves turn brown in the autumn?*

Because, when their power of decomposing the air declines, the *oxygen* absorbed in the carbonic acid gas, *lodges in the leaf*, imparting to it a red or brown colour.

1184. *Why do succulent fruits, such as gooseberries, plums, &c., taste acid?*

Because, in the formation of juices, a considerable amount of *oxygen* is absorbed, and the oxygen imparts acidity to the taste.

1185. *Why do ripe fruits taste sweet, and unripe fruits taste sour?*

Because the juices of the ripe fruit contain a large proportion of *sugar*, which in the unripe fruit has not been formed.

1186. *Why do some leaves turn yellow?*

Because they retain an excess of *nitrogen*. Leaves undergoing decay turn either yellow, red, crimson, or violet. Yellow is due to the excess of *nitrogen;* red and crimson to various proportions of *oxygen;* violet to a mixture of *carbon;* and green to *chlorophyl.*

1187. *Why do leaves fall off in the autumn?*

Because they have supplied for a season the natural wants of the tree. Every part has received nutrition through the spring and summer months; and the wants of the tree being supplied, the chief use of the leaf ceases, and it falls to the ground to decay, and enrich the soil.

1188. *Why do plants suffer from the smoke of cities?*

Because the smoke *injures the porous structure of the leaves,* and interferes with their free respiration.

LESSON LXI.

1189. *Why are vegetable productions so widely diffused?*

Because they everywhere form the *food of the animal creation.* Without them, neither man nor beast could exist. Even the flesh-eating animals are sustained by them, since they live by preying upon the bodies of vegetable-eaters.

They also enrich and beautify the earth. They present the most charming diversities of proportions and features. From the cowslip, the primrose, and the blue-bell of our childish days, to the broad oak under which we recline, while children gambol round us, they are all beautiful or sublime, and eminently useful in countless ways to man.

They spread a carpet over the surface of the earth; they cling to old ruins, and cover hard rocks, as though they would hide decay, and give warmth to the coldness of stone. In tropical climates they supply rich fruits full of cool and refreshing juices, and they spread out upon the crests of tall trees those broad leaves which shelter the native from the scorching heat of the sun.

They supply our dwellings with furniture of every kind, from the plain deal table, to the handsome cabinet of satin or rosewood; they afford rich perfumes to the toilette, and luscious fruits and wines to the desert; they charm the eye of the child in the daised field; they adorn the brow of the bride; they are laid in the coffin with the dead; and, as the cypress or the willow bend over our graves, they become the emblems of our grief.

1190. *What is mahogany?*

Mahogany is the wood of trees brought chiefly from South America and Spain. The finest kind is imported from St. Domingo, and an inferior kind from Honduras.

We all know the beauty of mahogany wood. But we do not all know that mahogany was first employed in the repair of some of Sir Walter Raleigh's ships at Trinidad in 1597. The discovery of the beauty of its grain for furniture and cabinet work was accidental. Dr. Gibbons, a physician of eminence, was building a house in King-street, Covent-garden; his brother, captain of a West Indiaman, had brought over some planks of mahogany as ballast, and he thought that the wood might be used up in his brother's building, but the carpenters found the wood too hard for their tools, and objected to use it. Mrs. Gibbons shortly afterwards wanted a small box made, so the doctor called upon his cabinet-maker, and ordered him to make a box out of some wood that lay in his garden. The cabinet-maker also complained that the wood was too hard. But the doctor insisted upon its being used, as he wished to preserve it as a memento of his brother. When the box was completed, its fine colour and polish attracted much attention; and he, therefore, ordered a bureau to be made of it. This was done, and it presented so fine an appearance that the cabinet-maker invited numerous persons to see it, before it was sent home. Among the visitors was her Grace the Duchess of Buckingham, who immediately begged some of the wood from Mr. Gibbons, and employed the cabinet-maker to make her a bureau also. Mahogany from this time became a fashionable wood, and the cabinet-maker, who at first objected to use it, made a great success by its introduction.

1191. *What is rosewood?*

Rosewood is the wood of a tree which grows in Brazil. It is, generally speaking, too dark for large articles of furniture, but is admirably adapted for smaller ones. It is expensive, and the hardness of the wood renders the cost of making articles of it very high.

"I am come up to the height of the mountains, to the sides of Lebanon, and will cut down the tall cedars thereof, and the choice fir trees thereof."—
II. KINGS XXIII.

Respecting the other woods used in the manufacture of furniture, we have nothing special to say, except of the oak—the emblem of our native land. This tree yields a most useful and durable wood, and as it not only defends our country by supplying our "wooden walls," but gives to us the floors of our houses, furnishes our good substantial tables, and comfortable arm-chairs, it will be well for us to know a few facts about this celebrated tree. It is said that there are no less than one hundred and fifty species of the oak. The importance of the growth of oaks may be gathered from the fact, that the building of a 70-gun ship would take forty acres of timber. The building of a 70-gun ship is estimated to cost about £70,000. Oak trees attain to the age of 1,000 years. The oak enlarges its circumference from 10½ inches to 12 inches in a year. The interior of a great oak at Allonville, in Normandy, has been converted into a place of worship. An oak at Kiddington, served as a village prison. A large oak at Salcey, was used as a cattle fold; and others have served as tanks, tombs, prisons, and dwelling-houses.

The *Mammoth tree*, which is exhibiting at the Crystal Palace, is one of the great wonders of the vegetable creation. It is the grand monarch of the Californian forest, inhabiting a solitary district on the elevated slopes of the Sierra Nevada, at 5,000 feet above the sea-level. From 80 to 90 trees exist, all within the circuit of a mile, and these varying from 250 to 320 feet in height, and from 10 to 20 feet in diameter. The bark is from 12 to 15 inches in thickness; the branchlets are somewhat pendent, and resemble those of cypress or juniper, and it has the cones of a pine. Of a tree felled in 1853, 21 feet of the bark from the lower part of the trunk were put in the natural form as a room, which would contain a piano, with seats for forty persons; and on one occasion 150 children were admitted. The tree is reputed to have been above 3,000 years old, that is to say, it must have been a little plant when Samson was slaying the Philistines. The portion of the tree exhibiting at the palace is 103 feet in height, and 32 feet in diameter at the base.

1192. *What is tea?*

Tea is the leaf of a shrub (*Thea Chinensis*). The plant usually grows to the height of from three to six feet, and resembles in appearance the well-known myrtle. It bears a blossom not unlike that of the common dog-rose. The climate most congenial to it is that between the 25th and 33rd degrees of latitude. The growth of good tea prevails chiefly in China, and is confined to a few provinces. The *green* and *black* teas are mere varieties, depending upon the culture, time of gathering, mode of drying, &c. *Coffee was used in this country before tea.* In 1664, it is recorded, the East India Company bought 2lb. 2oz. of coffee as a present for the king. In the year 1832, there were 101,687 licensed tea dealers in the United Kingdom. Green tea was first used in 1715. A dispute with America about the duty upon tea led to the American war, out of which arose American independence. The consumption of tea throughout the whole world is estimated at above 52,000,000 lbs, of which the consumption of Great Britain alone amounts to 30,000,000. (*See* 1225).

1193. *What is coffee?*

Coffee is the berry of the coffee plant, which was a native of that part of Arabia called Yemen, but it is now extensively cultivated in India, Java, the West Indies, Brazil, &c. (*See* 1224).

The first coffee-house in London was opened in 1652, under the following circumstances. A Turkey merchant named Edwards, having brought along with him from the Levant, some bags of coffee, and a Greek servant who was skilful in making it, his house was thronged with visitors to see and taste this new beverage. Being desirous to gratify his friends without putting himself to inconvenience, he allowed his servant to open a coffee-house, and to sell coffee publicly.

Here we have another illustration of the great results springing from trifling causes. Coffee soon became so extensively used that taxes were imposed upon it. In 1660 a duty of 4d. a gallon was imposed upon all coffee made and sold. Before 1732 the duty upon coffee was 2s. a pound; it was afterwards reduced to 1s. 6d., at which it yielded to the revenue, for many years, £10,000 per annum. The duty has been gradually reduced, and the consumption has gone on increasing, until at last above 25,000,000 of pounds are consumed annually! Fancy this great result springing from a "friendly coffee party" that assembled in the year 1652.

1194. *What is chocolate?*

It is a cake prepared from the cocoa-nut. The nut is first roasted like coffee, then it is reduced to powder and mixed with water, the paste is then put into moulds and hardened. The properties are very healthful, but its consumption is very insignificant, as compared with tea or coffee. The cocoa tree grows chiefly in the West Indies and South America.

1195. *What is cocoa?*

Cocoa is also a preparation from the seeds or beans of the cocoa tree. But the best form of cocoa for family use is to obtain the beans pure, as they are now commonly sold ready for use, and to break them and then grind them in a large coffee mill.

1196. *What is chicory?*

Chicory is the root of the common endive, dried and roasted as coffee, for which it is used as a substitute. Some persons prefer the flavour of chicory admixed with coffee. But very opposite opinions prevail respecting the qualities of chicory. We belive it to be perfectly healthful, and attribute the prejudice that prevails against it, to its having been used, from its cheapness, to adulterate coffee.

1197. *What is sugar?*

Sugar is a sweet granulated substance, which may be derived from many vegetable substances, but the chief source of which is the sugar cane. The other chief sources that supply it are the maple, beet-root, birch, parsnip, &c. It is extensively used all over the world. Sugar is supposed to have been known to the ancient Jews. It was found in the East Indies by Newcheus, Admiral of Alexander, 325 B.C. It was brought into Europe from Asia.

The art of sugar refining was first practised in England, in 1659, and sugar was first taxed by name by James II., 1685. Sugar is derived from the West Indies, Brazil, Surinam, Java, Mauritius, Bengal, Siam, the Isle de Bourbon, &c. &c. Before the introduction of sugar to this country, honey was the chief substance employed in making sweet dishes; and long after the introduction of sugar it was used only in the houses of the rich. The consumption in England in 1700 reached only 10,000 tons; in 1834 it had reached 180,000 tons. The English took possession of the West Indies in 1672, and in 1646 began to export sugar. In 1676 it is recorded that 400 vessels, averaging 150 tons, were employed in the sugar trade of Barbadoes. Jamaica was discovered by Columbus, and was occupied by the Spaniards, from whom it was taken by Cromwell, in 1656, and has since continued in our own possession. When it was conquered there were only three sugar plantations upon it. But they rapidly increased. Until the abolition of slavery in the West Indies, the production of sugar was almost exclusively limited to slave labour. (*See* 1226).

1198. *What is wheat?*

Wheat, rye, barley, oats, millet, and maize, all belong to the natural order of grain-bearing plants. They all grow in a similar manner, and all yield starch, gluten, and a certain amount of phosphates. They are commonly spoken of as *farinaceous foods.*

From the Sacred writings we learn that unleavened bread was common in the days of Abraham. In the earlier periods of our own history, people had no other method of making bread than by roasting corn, and beating it in mortars, then wetting it into a kind of coarse cake. In 1596, rye bread and oatmeal formed a considerable part of the diet of servants, even in great families. In the time of Charles the First, barley bread was the chief food of the people. In many parts of England it was more the custom to make bread at home then at present. In 1804, there was not a single public baker in Manchester. In France, when the use of yeast was first introduced, it was deemed by the faculty of medicine to be so injurious to health that its use was prohibited under the severest penalties.

"I clothed thee also with broidered work, and shod thee with badgers' skin, and I girded thee about with fine linen, and I covered thee with silk."—
EZEKIEL XVI.

Herault says that, during the siege of Paris by Henry the Fourth, a famine raged, and bread sold at a crown a pound. When this was consumed, the dried bones from the charnel house of the Holy Innocents were exhumed, and a kind of bread made therefrom. Bread-street, in London, was once a bread market. From the year 1266, it had been customary to regulate by law the price of bread in proportion to the price of wheat or flour at the time. This was called the assize of bread; but, in 1815, it was abolished. In the year 272 there was a famine in Britain so severe that people ate the bark of trees; forty thousand persons perished by famine in England in 310! In the year 450 there was a famine in Italy so dreadful that people eat their own children. A famine, commencing in England, Wales, and Scotland, in 954, lasted four years. A famine in England and France, in 1193, led to a pestilential fever, which lasted until 1195. In 1315 there was again a dreadful famine in England, during which people devoured the flesh of horses, dogs, cats, and vermin! In the year 1775, 16,000 people died of famine in the Cape de Verds. These are only a few of the remarkable famines that have occurred in the course of history. Let us thank God that we live in times of abundance when improved cultivation, the pursuit of industry, and the settlement of the laws, render such a calamity as a famine almost an impossibility.

1199. *What is cotton?*

Cotton is a species of vegetable wool, produced by the cotton shrub, called, botanically, *Gossypium herbaceum*, of which there are numerous varieties. It grows naturally in Asia, Africa, and America, and is cultivated largely for purposes of commerce.

The precise time when the cotton manufacture was introduced into England is unknown; but probably it was not before the 17th century. Since then, what wonderful advances have been made! The cotton trade and manufacture have become a vast source of British industry, and of commerce between nations. It was some years ago calculated that the cotton manufacture yielded to Great Britain one thousand millions sterling. The names of Hargreaves, Arkwright, Crompton, Cartwright, and others, have become immortalised by their inventions for the improvement of the manufacture of cotton fabrics. Little more than half a century has passed since the British cotton manufactory was in its infancy —now it engages many millions of capital—keeps millions of work people employed; freights thousands of ships that are ever crossing and re-crossing the seas; and binds nations together in ties of mutual interest. The present yearly value of cotton manufactures in Great Britain is estimated at £34,000,000. About £6,044,000 of the above sum is distributed yearly among working people as wages.

1200. *What is silk?*

Silk, though not directly a vegetable product, is, nevertheless, indirectly derived from the vegetable creation, since it is a thread spun by the silk-worm from matter which the worm derives from the *mulberry leaf.*

Silk is supplied by various parts of the world including China, the East

'And there was a man in Maon, whose possessions were in Carmel; and the man was very great, and he had three thousand sheep, and a thousand goats: and he was shearing his sheep in Carmel."—I SAMUEL XXV.

Indies, Turkey, &c., where the silk-worm has been found to thrive. The attempts that have been hitherto made to cultivate it in this country have proved unsuccessful. At Rome, in the time of Tiberius, a law passed the senate which, as well as prohibiting the wearing of massive gold jewels, also forbade the men to debase themselves by wearing silk. There was a time when silk was of the same value as gold—weight for weight—and it was thought to grow upon trees. It is recorded that silk mantles were worn by some noble ladies at a ball at Kenilworth Castle, 1286. It was first manufactured in England in 1604. In the reign of Elizabeth, the manufacture of silk in England made rapid strides. In 1666, there were 40 000 persons engaged in the silk trade. The silk throwsters of the metropolis were enrolled in a fellowship in 1562, and were incorporated in 1629. In 1635, a considerable impetus was given to the English silk manufactures. Louis the Fourteenth of France revoked the edict of Nantes. The edict of Nantes was promulgated by Henry the Fourth of France in 1598. It gave to the Protestants of France the free exercise of their religion. Louis the Fourteenth revoked this edict in 1685, and thereby drove the Protestants as refugees to England, Holland, and parts of Germany, where they established various manufactures. Many of these French refugees settled in Spitalfields, and there founded extensive manufactories, which soon rivalled those of their own country; and thus the intolerance of the king was justly punished. What important facts we see connected with the simple thread of the silk-worm!

1201. *What is wool?*

Wool is a kind of soft hair or coarse down, produced by various animals, but chiefly by sheep.

This is another of the useful productions of nature, for which we are indirectly indebted to the vegetable kingdom; for were it not for the rich pastures forming the green carpet of the earth, it would be impossible for man to keep large flocks of sheep for the production of wool. Wool, like the hair of most animals, completes its growth in a year, and then exhibits a tendency to fall off. For the production of wool in England and Wales it has been estimated that there are no less than 27,000,000 sheep and lambs; and, in Great Britain and Ireland, the total number is estimated at 32,000,000. Wool was not manufactured in any quantity in England until 1331, when the weaving of it was introduced by John Kempe and other artizans from Flanders. The exportation or non-exportation of wool has from time to time formed a vexed subject for legislators. Woollen clothes were made an article of commerce in the reign of Julius Cæsar. They were made in England prior to 1200. Blankets were first made in England in 1340. The art of dyeing wools was first introduced into England in 1608. The annual value of the raw material in wool is set down at £6,000,000; the wages of workmen engaged in the wool trade, £9,600,000, The number of people employed is said to be 500,000.

1202. *What is starch?*

Starch is one of the most useful products of the vegetable kingdom. As a rule, *a vegetable, if nutritious at all, is so according to the amount of starch which it contains.* It is most abundantly found in the seeds of plants, and especially in the *wheat* tribe.

It is also met with in the cellular tissues of plants, and especially in such underground stems as the *potatoe, carrot, turnip, &c.,* and the stems of the *sago-palm* fig, &c. It is also found in the *bark* of some trees.

1203. *Why is the horse chesnut, though containing a great quantity of starch, unfit for food?*

Because (like many other vegetable productions) it contains with the starch an *acrid juice,* which renders it unhealthy; and although the juice can be separated from the starch, the process is too expensive to be made generally available.

The starch which is used for domestic purposes is an artificial preparation, and does not properly represent the starch of nutrition. A better idea of it is afforded by *the meal of a flowery potatoe.* The starch used by laundresses is frequently prepared from diseased potatoes. This does not impair the quality of the starch, for the purposes of the laundress, and the reason why potatoes that are diseased are thus applied is, that it is one method of saving some part of their value. The finest kinds of starch are prepared from rice. It is prepared by breaking the pulp, and disengaging the starch from the cells; and it is then put through other processes to remove the fragments of the broken cells. But in the flowery meal of the potatoe, the starch cell may be seen entire.

CHAPTER LXII.

1204. *What are vegetable oils and fats?*

Vegetable oils and fats constitute, next to starch and sugar, the most important secretion of the vegetable creation. There are very few plants from which some amount of oil cannot be obtained; and those which are famed for yielding it owe their celebrity rather to the abundance that they yield, and the peculiar qualities of their oil, than to the secretion of oil being rare—for probably there is no plant without it.

Oil is most commonly found in seeds, as *rape-seed, linseed,* &c., but it is found also in leaves, as in the rose, sweet-briar, peppermint, &c., where its presence may be recognised by the distinguishing perfume; and it is also found in the wood of a few trees, such as the sassafras and the sandal-wood; the bark frequently yields an oily secretion.

The London and North Western Railway Company alone use about 50,000 gallons of oil yearly.

1205. *Why are fat and oil found most abundantly in the bodies of animals in cold climates?*

Because they contribute to keep the *bodies of animals warm,* not only by their non-conducting property *keeping in* the heat of the animals, but by supplying *carbon* abundantly to combine with *oxygen* during respiration, and thereby developing *animal heat.*

1206. *Why are oil and fat-forming trees found most abundantly in hot climates?*

Because, in hot countries, the formation of large quantities of fat in animal bodies would oppress living creatures with heat; fats and oils are, therefore, produced in those countries chiefly by vegetables, and are used externally by the Asiatics and Africans as an *external* unction for *cooling the skin,* and as *perfumes* which give inspiriting properties to the air, rendered oppressive by excess of heat.

1207. *Why are succulent fruits most abundant in tropical climates?*

Because they are rendered necessary in those climates by the *excessive heat,* and are found to have a most beneficial effect in cooling, purifying the blood of the inhabitants of tropical countries; while the grandeur of their foliage, and the richness of their flowers, are in perfect keeping with the intensity of light and heat, and serve, by throwing dense shades over the earth, to cool its surface, and to offer to living creatures a pleasant retreat from the rays of the burning sun.

The following sketch of *Botanical Geography* should be read attentively after the reader has gone through the whole of the Chapters of "Reasons." The technical terms employed in the course of the article are nearly all explained at 1212, and should be committed to memory at the commencement of the perusal. *Mimosa* means a sensitive plant; *concentric zones,* circular lines spreading from a centre; *arborescent,* resembling trees; *Gramineæ,* grass-like. The botanical names represent individual plants.

1208. When treating of the geographical distribution of vegetables, we have to mark the general arrangements indicated, and the agencies that have evidently operated in promoting the diffusion of floral tribes. Vegetation occurs over the whole globe, therefore, under the most opposite conditions. Plants flourish in the bosom of the ocean as well as on land, under the extremes of cold and heat in polar and equatorial regions, on the hardest rocks and the soft alluvium of the plains, amidst the perpetual snow of lofty mountains, and in springs at the temperature of boiling water, in situations never penetrated by the solar rays, as the dark vaults of caverns, and the walls of mines, as well as freely exposed to the influences of light and air. But these diverse circumstances have different species and genera. There is only one state which seems fatal to the existence of vegetable life—the entire absence of humidity.

1209. By species we understand so many individuals as intimately resemble each other in appearance and properties, and agree in all their permanent characters, which are founded in the immutable laws of creation. An established species may frequently exhibit new varieties, depending upon local and accidental causes, but these are imperfectly, or for a limited time, if at all, perpetuated.

1210. A genus comprises one or more species similar to each other, but essentially differing in formation, nature, and in many adventitious qualities from other plants. A tribe, family, group, or order, comprises several genera.

1211. The known number of species in the vegetable kingdom has been gradually enlarged by the progress of maritime and inland discovery; but owing to great districts of the globe not having yet been explored by the botanist, the interior of Africa, and Australia, with sections of America, Asia. and Oceanica, it is impossible to state the exact amount. The successive augmentation of the catalogue appears from the numbers below:

	Species.
Theophrastus	500
Pliny	1,000
Greek, Roman, and Arabian botanists	1,400
Bauhin	6,000
Linnæus	8,800
Persoon	27,000
Humboldt and Brown	38,000
De Candolle	56,000
Lindley	86,000
Hinds	89,000

1212. Vegetable forms are divided into three great classes which differ materially in their structure:—1. Cryptogamous plants—those which have no flowers, properly so called, mosses, lichens, fungi, and ferns: as distinguished from those which are phænogamous, or flower-bearing, to which the two following classes belong. 2. Endogenous plants, which have stems increasing from within, also called Monocotyledons, from having only one seed-lobe, as the numerous grasses, lilies, and the palm family. 3. Exogenous plants, which have stems growing by additions from without, also called Dicoteledons, from the seed consisting of two lobes, the most perfect, beautiful, and numerous class, embracing the forest trees, and most flowering shrubs and herds.

1213. The exogens furnish examples of gigantic size, and great longevity. In South America on the banks of the Atabapo, Humboldt measured a *Bombax caiba* more than 120 feet high, and 15 in diameter; and near Cumana, he found

He shall be like a tree planted by the rivers of water, that bringeth forth his fruit in season: his leaf also shall not wither; and whatsoever he doeth shall prosper."—PSALM I.

the *Zamang del Guayra*, a species of mimosa, the pendant branches of the hemispherical head having a circumference of upwards of 600 feet. The *Adansonia*, or baobab of Senegal, though attaining no great height, rarely more than fifty feet, has a trunk with a diameter sometimes amounting to 34 feet; while the *Pinus Lambertiana*, growing singly on the plains west of the Rocky Mountains, has been found 250 feet high, 60 feet in circumference at the base, 4½ feet in girth at the height of 190 feet, yielding cones 11 inches round, and 16 long. The *Ficus Indicus*, or banian tree, sending out shoots from its horizontal branches, which reaching the ground take root, and form new stems till a single tree multiplies almost to a forest, has been observed covering an area of 1700 square yards.

1214. From the number of concentric zones observed in a transverse section of the stems De Candolle advances proof of the following ages:

Elm	335	years.
Cypress	about 350	„
Cheirostemon	„ 400	„
Ivy	450	„
Larch	576	„
Orange	630	„
Olive	700	„
Oriental Plane	„ 720	„ and upwards.
Cedar of Lebanon	„ 800	„
Oak	810, 1080, 1500	„
Lime	1076; 1147	„
Yew	1214, 1458, 2588, 2880	„
Taxodium	4000 to 6000	„
Boabab	5150	„

1215. Admitting, with Professor Henslow, that De Candolle overrated the ages of these trees one-third, they are examples of extraordinary longevity. Yew trees upwards of 700 years old remain at Fountains Abbey, Yorkshire, as there is historic evidence of their existence in the year 1133. But a yew in the churchyard of Darley in the Dale, Derbyshire, is considered by Mr. Bowman as 2000 years old.

1216. The cryptogamous plants afford the most numerous examples of wide diffusion. A lichen indigenous in Cornwall, *sticta aurata*, is also a native of the West India Islands, Brazil, St Helena, and the Cape of Good Hope; while 38 lichens and 28 mosses are common to Great Britain and Australia, though the general vegetation of the two districts is remarkably discordant. Some species of endogenous plants are also widely distributed, the *Phleum alpinum* of Switzerland occurring without the slightest difference at the Strait of Magellan, and the quaking grasses of Europe in the interior of Southern Africa. But only in very few instances are the same species of exogenous plants met with in regions far apart from each other; and generally speaking, in passing from one country to another, we encounter a new flora; for if the same genera occur, the species are not identical, while in districts widely separated the genera are different.

1217. The cryptogamic plants, mosses, lichens, ferns, and fungi, are to the whole mass of phænogamic vegetation in the following proportions in different districts: Equatorial latitudes, 0 deg. to 10 deg.; on the plains, 1-25th, on the mountains, 1-5th; mean latitudes, 45 deg. to 52 deg. ½; high latitudes, 67 deg.

"To give unto them beauty for ashes, the oil of joy for mourning, the garment of praise for the spirit of heaviness; that they might be called Trees of righteousness, The planting of the Lord, that he might be glorified."—ISAIAH LXI.

70 deg., proportion about equal. Thus the proportion of the flowerless vegetation to the flowering increases from the equator to the poles. But the family of ferns, *filices*, viewed singly, forms an exception to this law, decreasing as we depart from equinoctial countries, being 1-20th in equatorial and 1-70th in mean latitudes, and not found at all in the high latitudes of the new world.

1218. In equinoctial and tropical countries, where a sufficient supply of moisture combines with the influence of light and heat, vegetation appears in all its magnitude and glory. Its lower orders, mosses, fungi, and confervæ, are very rare. The ferns are aborescent. Reeds ascend to the height of a hundred feet, and rigid grasses rise to forty. The forests are composed of majestic leafy evergreen trees bearing brilliant blossoms, their colours finely contrasting, scarcely any two standing together being of the same species. Enormous creepers climb their trunks; parasitical orchidæ hang in festoons from branch to branch, and augment the floral decoration with scarlet, purple, blue, rose, and golden dyes. Of plants used by man for food, or as luxuries, or for medicinal purposes, occurring in this region, rice, bananas, dates, cocoa, cacao, bread-fruit, coffee, tea, sugar, vanilla, Peruvian bark, pepper, cinnamon, cloves, and nutmegs, are either characteristic of it as principally cultivated within its limits, or entirely confined to them.

1219. Rice *(Oryza-sativa)*, the chief food of, perhaps, a third of the human race, is cultivated beyond the tropics, but principally within them, only where there is a plentiful supply of water. It has never been found wild; its native country is unknown; but probably southern Asia.

1220. Bananas, or plantains *(Musa sapientum et paradisiaca)*, are cultivated in intertropical Asia, Africa, and America. The latter species occur in Syria. The banana is not known in an uncultivated state. Its produce is enormous, estimated to be on the same space of ground to that of wheat, as 133 to 1, and to that of potatoes as 44 to 1.

1221. Dates *(Phœnix dactylifera)*, and cocoa *(Cocos nucifera)*, belonging to the family *Palmæ*. The palms, remarkable for their elegant forms and importance to man, contribute more than any other trees to impress upon the vegetation of tropical and equinoctial countries its peculiar physiognomy. The date palm is a native of northern Africa, and is so abundant between the Barbary states and the Sahara, that the district has been named Biledulgerid, the land of dates. As the desert is approached, the only objects that break the monotony of the landscape are the date palm, and the tent of the Arab. It accompanies the margin of the mighty desert in all its sinuosities from the shores of the Atlantic to the confines of Persia, and is the only vegetable affording subsistence to man that can grow in such an arid situation. The annual produce of an individual is from 150 to 260lbs. weight of fruit. The cocoa palm furnishes annually about a hundred cocoa-nuts. It is spread throughout the torrid zone; but occurs most abundantly in the islands of the Indian archipelago. The family of palms is supposed to contain a thousand species, some of large size, forming extensive forests.

1222. Cacao *(Theobrama cacao)*, from the seeds of which chocolate is prepared, grows wild in central America, and is also extensively cultivated in Mexico, Guatamala, and on the coast of Cumana.

1223. Bread-fruit tree *(Artocarpus incisa)*, a native of the South Sea Islands, and Indian archipelago, grows also in Southern Asia, and has been introduced

"And they returned and prepared spices and ointments; and rested the Sabbath-day, according to the commandment."—LUKE XXIV.

into the tropical parts of America; but the fruit is not equal to the banana as an article of human food.

1224. Coffee *(Coffea Arabica)*. The bush has probably for its native region the Ethiopian Highlands, from whence it was taken in the fifteenth century to the Highlands of Yemen, the southern part of the Arabian peninsula. It has been introduced, and is now extensively cultivated in British India, Java, Ceylon, the Mauritius, Brazil, and the West Indies, but the quality is inferior, which makes the climate of the Mocha coffee district of importance, as peculiarly favourable to the plant. It grows there on hills described by Niebuhr as being soaked with rain every day from the beginning of June to the end of September, which is carefully collected for the purpose of irrigation during the dry season. Forskhal gives the following temperatures in the district:

Beit el Fakih ...	March	16,	7 A.M.	76 deg.	1 P.M.	95 deg.	10 P.M.	81 deg.
„ ..	„	18,	„	77	„	95	„	81
Hodeida	„	18,	„	72	„	92¾	„	78
Bulgosa, a village in the hills ...	„	20,	„	69½	„	85½	„	73

1225. Tea *(Thea Chinensis)*. The plant is indigenous in China, Japan, and Upper Assam. In the latter country, it has recently been found in a wild state, and is in process there of extensive cultivation. As the plant is hardy, its culture has very lately been attempted in the South of France, and apparently with complete success. A similar experiment on the burning plains of Algeria completely failed, all the plants being killed by the heat, notwithstanding every precaution. Tea was first introduced into Europe by the Dutch in 1666. The leaves of the coffee-plant have long been used as a substitute for tea, by the lower classes in Java and Sumatra; and recently, Professor Blume, of Leyden, exhibited samples of tea prepared from coffee-leaves, agreeing entirely in appearance, odour, and taste, with the genuine Chinese production.

1226. Sugar-cane *(Saccharum officinaram)*, a species of *Gramineæ*, occurs to some extent without the tropics, having been cultivated centuries ago in Europe, as at present scantily in the South of Spain. But it properly belongs to the torrid zone, and has for its principal districts, the Southern United States, the West Indies, Venezuela, Brazil, the Mauritius, British India, China, the Sunda and Philippine Islands. The plant was found wild in several parts of America on the discovery of that continent, and occurs in a wild state on many of the islands of the Pacific.

1227. Vanilla *(Vanilla aromatica)*, the fruit of which forms the well-known aromatic, grows wild principally in Mexico.

1228. Peruvian bark (*Cinchona officinalis*), a forest tree, of which there are several species, furnishing the valuable medicine so called. It is exclusively confined to South America, and grows chiefly on the Andes of Loxa and Venezuela.

1229. Pepper (*Piper nigrum)* belongs exclusively to the Malabar coast, where it has been found wild, Sumatra, which produces the greatest quantity, Borneo, the Malay peninsula, and Siam. Other species of *Piperaceæ* occur in tropical America.

1230. Cinnamon (*Laurus Cinnamomum)*, a small tree yielding the aromatic bark, is found native only in the island of Ceylon; but another species occurs in Cochin China.

"I am the true vine, and my Father is the husbandman."—JOHN XV.

1231. Clove (*Myrtus caryophyllus*), an evergreen small tree, the dried flower-buds of which form the celebrated aromatic, grows naturally in the Moluccas, whence it has been conveyed to other tropical districts. The island of Amboyna, one of that group, is the principal seat of its cultivation. The lowest temperature there is 72 degs.; the mean temperature of the year 82 degs.

1232. Nutmeg *(Myrstica moschata)* grows naturally in several islands of the eastern archipelago, but is principally cultivated in the Banda Isles.

Tropical families and forms successively vanish with an increase of distance from the equator, and new phases of vegetation mark the transition from hot to temperate climates. Vividly green meadows, abounding with tender herbs, replace the tall rigid grasses which form the impenetrable jungle; and instead of forests composed of towering evergreen trees, woods of the deciduous class appear, which cast their leaves in winter, and hybernate in the colder season, the oak, ash, elm, maple, beech, lime, alder, birch, and sycamore. The cultivation of the vine becomes characteristic, with the perfection of the cereal grasses, and a larger proportion of herbaceous annuals and cryptogamic plants.

1233. The vine *(Vitis vinifera)* is less impatient of a cold winter than a cool summer. Hence its northern limit, which coincides with lat. 47 deg. 30 min. on the west coast of France, rises in the interior, where, though the winters are colder, the summers are warmer, to lat. 49 degs., cuts the Rhine at Coblentz in lat. 50 deg. 20 min., and ascends to 52 deg. 31 min. in Germany.

1234. Receding further from the equator, magnificent forests of the fir and pine tribe prevail, as in the central parts of Russia, on the southern shores of the Baltic, in Scandinavia, and North America. But some of the cereals are no longer cultivatable, and several timber-trees common to the temperate zone do not reach its northern limits. Gradually all ligneous vegetation disappears entirely as higher latitudes are approached, the woods having first dwindled to mere dwarfs in struggling with the elements, hostile to that state which nature destined them to assume. The limit of the forests is a sinuous line running along the extreme north of the old world; and extending from Hudson's Bay, lat. 60 deg., to the Mackenzie River, lat. 68 deg., and thence to Behring's Strait. The dwarf birch *(Betula nana)*, a mere bush, is the last tree found on drawing near the eternal snow of the pole. At the island of Hammerfest, lat. 70 deg. 40 min., near the North Cape, it rises to about the height of a man, in sheltered hollows between the mountains, its lower branches trailing on the ground, affording a shelter to the ptarmigan. In the polar zone, some low flowering annuals, saxifrages, ranunculi, gentians, chickweeds, and others, flourish during the brief ardent summer; a few perennials also accommodate themselves to the rigorous climate by spreading laterally, never rising higher than four or five inches from the ground; till finally no development of vegetable life is met with, but lichens, and the microscopic forms that colour the snow.

1235. In Europe, wheat ceases with a line connecting Inverness in Scotland, lat. 58 deg., Drontheim in Norway, lat. 64 deg., and Petersburgh in Russia lat. 60 deg. 15 min. Oats reach a somewhat higher latitude. Barley and rye ascend to lat. 70 deg., but require a favourable aspect and season to produce a crop.

1236. The northern limit of the growth of oak, lat. 61 deg., falls short of that of wheat. The oak makes a singular leap at the confines of Europe and Asia, disappearing towards the Ural mountains. This is the casea lso with the wild-nut and apple. The oak and the wild-nut, however, re-appear suddenly in

Eastern Asia, on the banks of the Argoun and the Amour; and the apple occurs again in the Aleutian Isles.

1237. The following are the northern limits of several trees in Scandinavia:

	Lat.
Beech, *Fagus silvatica*	60 deg. 0 min.
Hard Oak, *Quercus robur* . .	61 " 0 "
Common Elm, *Ulmus campestris* .	61 " 0 "
Common Lime, *Tilia communis* .	61 " 0 "
Common Ash, *Fraxinus excelsior* .	62 " 0 "
Fruit trees	63 " 0 "
Hazel, *Corylus avellana*	64 " 0 "
Spruce Fir, *Abies excelsa*	67 " 40 "
Service Tree, *Sorbus aucuparia* .	70 " 0 "
Scotch Fir, *Pinus silvestris* . .	70 " 0 "
White Birch, *Betula alba* . .	70 " 40 "
Dwarf Birch, *Betula nana* . .	71 " 0 "

1238. Thus distinct vegetable regions are observed on passing from south to north through different climatic zones, defined as to their limits by the isothermal curves, and not by the parallels of latitude. Similar changes of vegetation mark a perpendicular transit through varying climates. A succession of plants appear on the tropical mountains which rise above the snow line, corresponding to those which are encountered in mean and high latitudes. The higher we ascend, the more does the number of the phænogamic class diminish in proportion to the cryptogamic, till only members of the latter family are found, whose further progress upward is arrested by the everlasting snow. The last lichen met with by Saussure on Mont Blanc, *Silene acaulis*, was also observed by M. Brevais in the neighbourhood of Bosekop, lat. 69 deg. 58 min. where it was vegetating on the sea-shore, shaded by the last pines of Europe.

1239. Isolated mountains display to the best advantage the effect of climatic change of vegetation.

1240. Etna is divided into three great regions: *La Regione Culta* . . fertile region; *La Regione Sylvosa*, or woody region; *La Regione Deserta*, the bare or desert region. But each of these is susceptible of sub-divisions, defined by the presence of certain families of plants, forming seven botanical zones.

1. The sub-tropical zone, which does not rise more than 100 feet above the evel of the sea, is characterised by the palm, banana, Indian fig, sugar-cane, varities of mimosa and accacia, which with us are only found in conservatories.

2. The hilly zone rises about 2,000 feet, characterised by the orange, lemon, shaddock, maize, cotton, and grape plants.

3. The woody zone lies between the height of 2,000 and 4,000 feet, where the cork-tree flourishes, several kinds of oak, the maple, and enormous chesnuts.

4. The zone between the height of 4,000 and 6,000 feet is distinguished by the beech, Scotch fir, birch, and, among small plants, by clover, sandwort, chickweed, dock, and plantain.

5. The sub-alpine zone, between the elevation of 6,000 and 7,500 feet, produces the barberry, soap-wort, toad-flax, and juniper.

6. The zone between 7,500 and 9,000 feet, has almost all the plants of the preceding, with the fleshy and jagged groundsel.

"In the mountain of the height of Israel will I plant it; and it shall bring forth boughs, and bear fruit, and be a goodly cedar: and under it shall dwell all fowl of every wing; in the shadow of the branches thereof shall they dwell."—
EZEKIEL XVII.

7. The narrow zone between 9,000 and 9,200 feet, only produces a few lichens, beyond which, there is complete sterility.

1241. The Peak of Teneriffe exhibits five botanical districts, thus distinguished by Von Buch:

1. The region of Africa forms, 0—1,248 feet, comprising palms, bananas, the sugar-cane, various species of arborescent *Euphorbiæ*, *Mesembryanthema*, the *Dracæna*, and other plants, whose naked and tortuous trunks, succulent leaves, and bluish-green tints, are distinctive of the vegetation of Africa.

2. Region of Vines and Cereals, 1,248—2,748 feet, comprising also the olive, and the fruit-trees of Europe.

3. Region of Laurels, 2,748—4,350 feet, including lauri of four species, the wild olive, an oak, the iron-tree, the arbutus, and other evergreens. The ivy of the Canaries and various twining shrubs cover the trunks of the trees, and numerous species of fern occur, with beautiful flowering plants.

4. Region of the Pines, 4,350—6,270, characterised by a vast forest of trees resembling the Scotch fur, intermixed with juniper.

5. Region of the Retama, 6,270—11,061 feet, a species of broom, which forms oases in the midst of a desert of ashes, ornamented with fragrant flowers, and furnishing food to the goats which run wild on the Peak. A few gramineous and cryptogamic plants are observed higher, but the summit is entirely destitute of vegetation.

1242. There are many plants which can accommodate themselves to the most diverse climates and localities; and therefore ascend from the plains close to the boundary of vegetable life on the highest mountains. But it is the general law in these cases for such plants to be singularly modified in appearance and anatomical structure as they ascend. The spring gentian, *Gentiana verna*, is one of the exceptions, which Raymond found unaltered at all heights in the Pyrenees.

1243. Trees, plants, and bushes, of humbler growth, which occur on the plains and at great heights, are usually much smaller in the latter situation. The leaves, and everything green about them, dwindle with the increased elevation; and the pure, well defined green is exchanged for an ill-defined light yellow. Singular enough, those parts which seem most capable of resisting cold, as the leaves and stalks, are uniformly subjected to a diminution of their vital functions; while the flowers remain of the same size, are never deformed, and become more dense and richer in their colours. While the *Myosotis silvestris* becomes stunted, its flowers assume an intense blue—the admiration of the traveller. The flowers of the pale primrose have a much deeper colour on the top of the Faulhorn, while the plant itself is much smaller than its congener on the Swiss plains. The observations of M. Parrot, among others, are to this effect on the flora of the Caucasus, of Ararat, the Swiss and Italian Alps, and the Pyrenees. The arctic flora is similarly distinguished.

1244. The preceding references to different climatic states are, however, perfectly inadequate to explain the phenomena of vegetable distribution. While an analogy is often observable between the plants of different regions under corresponding circumstances of latitude, elevation, and soil, the species are generally found to be different; and usually the botanical character of countries not widely apart from each other, is totally different, though under the same parallels.

1245. Some plants are entirely confined to one side of our planet. The beautiful genus *Erica*, or heath, of which there are upwards of 300 species, occurs with breaks over a narrow surface, extending from a high northern latitude to the Cape of Good Hope. But the whole continent of America does not contain a single native specimen; nor has a *Pœnia* been found in it, except a solitary one to the west of the Rocky Mountains. On the other hand, the New World contains many families, as the *Cacti*, which are not found naturally in the Old.

1246. Some plants occur in a single specific locality, frequently a contracted area, and nowhere else. The beautiful *Disa grandiflora* is limited to a spot on the top of the Table Mountain at the Cape; and the celebrated cedar of Lebanon appears to be restricted in its spontaneous growth to the Syrian mountains. The small island of St. Helena has an indigenous flora, with a few exceptions different from that of the rest of the globe.

1247. Mountain chains of no great width very commonly divide a totally distinct botany. There is a marked difference in the vegetation of the Chilian and opposite side of the Andes, though the climate as well as the soil is nearly the same, and the difference of longitude very trifling. In North America, two completely different classes of vegetation appear on the two sides of the Rocky Mountains. A variety of oaks, palms, magnolias, azaleas, and magnificent rhododendrons occur on the eastern side, all of which are unkn own on the western, the region of the giant pine.

1248. The distinct vegetation possessed by various parts of the globe, has led to its division into botanical kingdoms or phyto-geographical regions, named in general after the genera that are either peculiar to them, or predominant in them. The arrangement of M. Schouw, which is usually adopted, discriminates twenty-five great provinces of characteristic vegetation upon the surface of the earth.

In constituting any portion of the globe into a phyto-geographical region, M. Schouw has proceeded upon the following principles:—1. That at least one-half of the species should be indigenous in it. 2. That a-quarter of the genera should also be peculiar to it, or at least should have a decided maximum. 3. That individual families of plants should either be exclusively confined to the region, or have their maxima there.

1249. The phenomena of botanical geography, and the facts of geology, are mutually illustrative. The existing dry land having been upheaved above the waters at different epochs, it may be reasonably inferred that each portion on its emergence received a vegetable creation in harmony with its position. The ultimate constitution of the general surface into different botanical kingdoms would hence follow, each of which has preserved its primitive features, while adjoining, and even far distant foci, have to some extent intermingled their respective products, under control of the natural agencies of diffusion.

1250. The agents that involuntarily officiate in the diffusion of vegetable products are the atmosphere, the waters, and many animals.

1. The impulsion of the atmosphere in its calmest state, is quite sufficient to transport to considerable distances seeds furnished with downy appendages or winglets, as is the case with many plants, with the minute sporules of

"He shall come down like rain upon the mown grass, as showers that water the earth."—PSALM LXXII.

cryptogamia, which are light as the finest powder. When ordinary breezes convey the sand-dust of the Sahara a thousand miles or more from the desert, it may be conceived that seeds, which are comparatively heavy, are borne far from home by the hurricane. Two Jamaica lichens, which had never been seen in France before, were found by De Candolle growing on the coast of Brittany, the offspring of sporules which had been swept over the Atlantic.

2. The mountain torrent washes down into the valley the seeds that have accidentally fallen into it, or have been swept away by its overflows; and hence the plants of the High Alps occur on the plains of Switzerland, which are entirely wanting in France and Germany. Rivers answer the same purpose more extensively, and also the oceanic currents. The nicker-tree, one of the leguminous tribe, has been raised from seed borne across the Atlantic by the Gulf stream.

3. Animals of the sheep and goat kinds, with the horse, deer, buffalo, and others, widely disperse several species of plants, the seeds of which, furnished with an apparatus of barbs and hooks, adhere to their coating. Seeds also of various kinds pass through the digestive organs of birds, uninjured as to their vitality. The little squirrel buries the acorn in the ground for winter provender, and sows an oak, if prevented from returning to the spot.

1251. Plants capable of extended naturalisation, and serviceable as articles of food or luxury, have been widely disseminated by the human race in their migrations. The cerealia afford a striking example. These important grasses known to the ancients, wheat, barley, oats, and rye, were the gifts of the Old World to the New. They are also importations into Europe; but the loose reports of the ancients, and the diligent researches of the moderns, alike leave us in ignorance of their native seat. Probability points to the conclusion that they have spread from the neighbourhood of the great rivers of Western Asia, the primitive location of the human family; and it is not impossible that in that imperfectly explored district, or further east on the Tartarian table-land, some of the cereals may yet be found growing spontaneously. The first wheat sown in North America, consisted of a few grains accidentally found by a negro slave of Cortes, among the rice taken for the support of his army. In South America the first wheat was brought to Lima by one of the early colonists, a Spanish lady, Maria d'Escobar. An ecclesiastic, Jose Rixi, was the first to sow it in the neighbourhood of Quito.

1252. Maize, or Indian corn *(Zea mays)*, has been dispersed in the Old World from the New; and also a more important product, the potato (*Solanum tuberosum)*, the use of which now extends from the extremity of Africa to Lapland. In Chili, the native country of the plant, it occurs at present in a wild state. The Spaniards imported it into Spain, and from thence it was communicated to Italy. It was first made known in England at a subsequent period from Virginia, having been received there from the Spanish colonists in South America, as it is not a native of intervening Mexico.

1253. The grape-vine, so extensively spread over Europe, is probably not indigenous in any part of it. It chiefly owes its diffusion there to the Romans, who received it from the Greeks, to whom it most likely immediately came from the country between the Black and Caspian Seas. The Romans introduced most of the finer European fruit-trees, some from Africa, as the pomegranate, but the great majority from Western Asia, as the orange, fig cherry, peach, apricot, apple, and pear. A variety of the plum. the damson, or damascene, came from the neighbourhood of Damascus during the Crusades. The name of the damask-rose points to the importation of the plant from the same quarter into Europe.

The ocean as well as the land has different botanical regions; and changes of the vegetation are observed with the depth analogous to the variations of terrestrial plants with the height. Marine vegetation seems to have its vertical extent determined by the range of light in water, which varies with the power of the sun and the transparency of the water.

CHAPTER LXIII

1254. *What are vegetable gums?*

Vegetable gums are secretions of plants which are generally *soluble in water*, and which subserve various useful purposes. *Gum Arabic* is one of the most important of this class of vegetable productions.

Gutta-percha is an invaluable substance lately added to the list of known vegetable productions. It is obtained by cutting the bark of trees of the class called *Sapotacea.* Its proper name is gutta Pulo Percha, gutta meaning gum, and Pulo Percha is the island whence it is obtained. But gutta-percha is not, strictly speaking, a gum.

India-rubber is also a vegetable secretion, improperly called elastic gum. It is obtained from the milky juice of various trees and plants, especially from the syringe tree, of Cayenne.

1255. *What are vegetable resins?*

Vegetable resins are derived from the secretions of plants, and are generally distinguished from gums by being *insoluble in water*, but being soluble in spirits.

When one of these substances is soluble in either water or spirits it is called a *gum-resin.*

1256. *What are vegetable acids?*

Vegetable acids are chiefly obtained from *fruit*; but also abundantly from *wood*, by distillation.

"Thou art the God that doest wonders."—PSALM LXXVII.

1257. *What is tannin?*

Tannin is a vegetable production, obtained chiefly from the oak-bark, and from a variety of other vegetable sources. It possesses the peculiar chemical property which renders it valuable in tanning leather.

1258. *What is opium?*

Opium is the produce of the *poppy*, and is obtained from the seed.

1259. *What are vegetable dyes?*

Vegetable dyes are the various colours derived from the secretions of plants, such as *indigo, madder, logwood, alkanet-root, &c.*

1260. *What is silica?*

Silica is a mineral substance, commonly known as *flint*; and it is one of the wonders of the vegetable tribes, that, although flint is so indestructible that the strongest chemical aid is required for its solution, plants possess the power of *dissolving and secreting* it. Even so delicate a structure as the wheat straw dissolves silica, and every stalk of wheat is covered with a perfect, but inconceivably thin coating of this substance.

Amid all the wonders of nature which we have had occasion to explain, there is none more startling than that which reveals to our knowledge the fact that a flint stone consists of the mineralised bodies of animals, just as coal consists of masses of mineralised vegetable matter. The animals are believed to have been infusorial animalculæ, coated with silicous shells, as the wheat straw of to-day is clothed with a glassy covering of silica. The skeletons of animalculæ which compose flint may be brought under microscopic examination. Geologists have some difficulty in determining their opinions respecting the relation which these animalculæ bear to the flint stones in which they are found. Whether the animalculæ, in dense masses, form the flint; or whether the flint merely supplies a sepulchre to the countless millions of creatures that, ages ago, enjoyed each a separate and conscious existence, is a problem that may never be solved. And what a problem! The buried plant being disentombed, after having lain for ages in the bowels of the earth, gives us light and warmth; and the animalcule, after a sleep of ages, dissolves into the sap of a plant, and wraps the coat it wore, probably "in the beginning, when God created the heavens and the earth, and when the earth first brought forth living creatures," around the slender stalk of waving corn!

1261. *Why is silica diffused over the stems of wheat, grasses, canes, &c.?*

Because it affords strength, density, and durability, to structures

"For in this mountain shall the hand of the Lord rest, and Moab shall be trodden down under him, even as straw is trodden down for the dunghill."—
ISAIAH XXV.

that are very light, and which, but for this beautiful provision, would be exceedingly perishable.

1262. *Why is guano a productive manure?*

Because it contains, with other suitable elements, an abundance of the *silicous skeletons of animalculæ.*

1263. *Why does a wheat-crop greatly exhaust the soil?*

Because, as well as the *carbon, and the salts,* which form the straw and the grain, it draws off from the soil a great amount of *silica.*

1264. *Why is straw frequently used as a manure?*

Because it gives back, with other substances, a *considerable proportion of silica,* in that form which adapts it to the use of the succeeding crop.

1265. *Why is the structure of herbaceous plants less consolidated than that of woody plants?*

Because, for the most part, herbaceous plants last only *a single year;* they, therefore, do not require the enduring qualities of plants that have to sustain the influences of the elements for a succession of seasons.

1266. *Why are the stalks of plants of light structure generally cylindrical?*

Because the cylindrical form is stronger than any other; *a hollow cylinder,* with moderately thick walls, *is stronger than a solid rod,* containing the same amount of material.

1267. *Why do the stalks of plants become hollow?*

Because the parallel and perpendicular fibres of the stalk are developed *more rapidly than the horizontal.* The growth of the plant, therefore, consists of a kind of *divergence from the centre.*

1268 *Why are the stomata, or pores of leaves, generally placed on their under surface?*

Because, being placed on the under surface, they are *shaded* from the action of the *sun's rays,* and so carry on the function of respiration more actively than if subjected to direct heat; they are also protected from the injurious *effects of dust;* and are moistened by *evaporation from the earth's surface.*

1269. *Why have plants a formation of pith in their centre?*

The pith is the chief organ of nutriment, especially in the young plant. It is the structure which first conveys fluids to, and receives them from, the newly-formed leaf. It communicates with every branch, leaf, bud, and flower; and also with the bark, through the *medullary rays,* which radiate horizontally from the centre of the plant. It is the centre of the movements of the sap which occur in the horizontal vessels; and it holds an important influence over the life of the plant.

1270. *Why are trees covered with bark?*

Because the bark serves to protect the woody structure, and also to give a passage to the descending sap which flows abundantly in the spring, and out of which the woody fibre is formed. It is also, from its peculiar nature, well fitted to endure the changes of the seasons for many years; and from its non-conducting properties it serves to maintain the equal temperature of the vital parts of the tree.

1271. *What is cork?*

Cork is the bark of a description of *oak-tree,* which grows in great abundance in Spain, Italy, and France.

1272. *Why does the cork-tree release its own bark?*

Because it possesses a bark which is exceedingly *useful to man;* and it seems, therefore, to have been the design of providence that the tree should cast it off, to be applied to the wants of the human family; for the cork-tree does not discharge its bark by the mere cracking, or exfoliation, of its substance; the tree retains the bark for a number of years, until it has attained that consistency and thickness which renders it useful, and then the tree forms within the bark a series of tabular cells, which *cut off the connection of the bark with the internal structure,* after which it peels off in large sheets.

"And all the trees of the field shall know that I the Lord have brought down the high tree, have exalted the low tree, have dried up the green tree, and have made the dry tree to flourish: I the Lord have spoken, and have done it.—EZEK. XVII.

Man assists this evident intention of nature, by slitting the bark from the top of the tree to its base; but even were this not done, the bark would be cast off by the tree itself

Another proof of design in this useful adaptation of the cork-tree is to be found in the fact, that it thrives under treatment that would destroy other trees. The cork-tree will endure the barking process for *seven or eight successive years.*

CHAPTER LXIV.

1273. *Why are there curious markings in walnut, mahogany, rose-wood, satin-wood, &c.?*

Because those markings are produced by the various *structure of the vessels* by which the wood is formed; and by successive zones of wood, which indicate the periods of growth.

The inclosure of zone within zone is owing to the mode in which the wood is produced, and the position in which it is deposited. Wood is formed by the leaves during the growing season, and passes down towards the root between the bark and the wood of the previous year (if any), or in the position in which cambium is effused; and, as the leaves more or less surround the whole stem, the new layer at length completes a zone, and perfectly encloses the wood of all former years. This is the explanation of the term *exogenous*, which is derived from two words signifying to grow outwardly, for the stem increases in thickness by successive layers on the outer side of the previously-formed wood. That this is the mode of growth has been abundantly proved by experiment, and demonstrated by accidental discoveries. Thus, if a plate of metal be inserted between the bark and wood, it will, in progress of time, become inclosed by the new wood which has overlaid them. So in like manner if letters be cut deeply through the bark and into the wood, the spaces will not be filled up from the bottom, but may be seen in subsequent years overlaid by new wood. A statement appeared in a daily paper, during the past year, to the effect that in cutting down a tree a cat had been discovered inclosed in the wood of the trunk, These facts prove that the wood is applied from without. Again, if a branch be stripped of its leaves down to a certain point, it will not grow above that point; and so, in like manner, if branches be stripped from one side of a tree, the tree will not grow on that side. If a circle of bark be removed from a branch above and also below a leaf, it will be found that increase of size will occur below, but not above that bud; and so, likewise, whenever a ring of bark is removed from a tree, the new woody fibre will not proceed from the lower but from the upper edge.—*Orr's Circle of the Sciences.*

"And when he saw a fig tree in the way, he came to it, and found nothing thereon, but leaves only, and said unto it, Let no frut grow on thee henceforward for ever. And presently the tree withered away."—MATTHEW XXI.

1274. *Why have trees with large trunks a great number of leafy branches?*

Because it is *by the leaves* that the secretion is formed which supplies the *woody fibre.* The number of leaves on a tree, therefore, generally bears a relation to the size of its trunk, and the number of its branches.

1275. *Why have poplar-trees comparatively few branches and leaves?*

Because their trunks are comparatively *small,* although they grow to a great height.

1276. *Why had the mammoth-tree comparatively few leaves in relation to the immense size of its bark?*

Because the woody texture of this tree *(Wellingtonea gigantea)* is *exceedingly light and porous.* It is, in fact, lighter than cork, and, therefore, requires less leaf-produce in its formation.

1277. *Why have oak-trees an abundance of leaves?*

Because their wood is *so dense* that they require a larger amount of the wood-forming secretion which is supplied by the leaves.

1278. *Why are the trunks of trees round?*

Because, generally speaking, the leaves are distributed upon branches around the trees in every direction. They consequently send down the wood-forming principle on all sides. When a trunk is unduly developed on one side, it may generally be traced to the unequal distribution of the branches.

1279. *What are exogenous stems?*

Exogenous stems are those that grow by the addition of wood *on their outer surface,* underneath the bark.

1280. *What are endogenous stems?*

Endogenous stems are those that *grow inwardly,* from the centre. Trees of this class, of which palms are the best example, are almost peculiar to tropical climates.

1281. *Why do endogenous stems chiefly abound in tropical climates?*

Because, probably, the excessive heat of those climates would interfere with the *formation of wood from the sap* upon the outer surface.

The vascular structure of endogenous stems lying more abundantly towards their centre, tends to *conserve the juices* which in hot climates are so highly valued. Palm-wine is a delicious and cooling beverage, and is procured from various kinds of palms, but especially from the cocoa-nut palm. Even the fresh sap is very refreshing. The juice is procured by cutting the tree in the upper part, and attaching a vessel to the opening, to receive the sap. Its flow is increased by cutting off a slice of the wood daily.

1282. *Why have endogenous stems no bark?*

Because, one of the chief functions of the bark in exogenous trees, is to *protect the sap* from which the wood is formed on the outward surface; and as there is no such external flow of sap in endogenous trees, the bark is *unnecessary to them*, and is therefore withheld. They are furnished instead with a thin cuticle.

1283. *Why do endogenous stems grow to a great height?*

Because, as the stem grows from the centre, it soon reaches that limit of diameter *which its vascular structure is calculated to support;* and, therefore, the wood-forming sap is deposited chiefly at the top of the stem, causing it to grow to a considerable height.

1284. *Why do the various vegetable fruits ripen in succession?*

Because the Author of Nature has thus arranged its economy, *in order that the wants of living creatures may be adequately provided for*. Some vegetable productions arrive at their perfection in the spring; others in summer; and others in autumn. Among the latter are many that require to come slowly to maturity after they are gathered; by these the winter season is provided for, and a surplus of the winter stock goes to supply the natural deficiency of spring.

1285. *Why, when seeds are sown, and germination begins, does the leaf-germ seek the light, and the root-germ grow down into the earth?*

Because the Creator has endowed every single seed with a

vital instinct which governs its development. The rootlet could more easily grow upward than downward, because of the looser earth, and of the exciting influences of light and moisture. Yet it takes the contrary course, leaving the leaf-germ to come up to meet the sun-light, and to send down to the stem and roots, the matter needed for their growth.

Frequently, indeed, when seeds are thrown into the earth, their natural position is reversed, and when the germs first start from the seed, the *root-germ* is directed *upward* and the *leaf-germ downward.* What then occurs? They each turn, and, in doing so, frequently cross each other. Each goes to its particular duty—the duty that God appointed.

CHAPTER LXV.

1286. *Why are the seeds of plants indigestible?*

Because they are encased in a hard covering upon which the gastric juice of animals takes no effect. This provision has been made by the Creator, *for the preservation of seeds,* the productions of which are so essential to animal life.

The gastric juice can dissolve any other part of the plant, even the woody fibre, and yet upon the *seed* it takes no effect. When, however, the seed is *crushed,* and, thereby, the vital principle destroyed, so that no plant can spring from it, the gastric juice acts upon it, and it is soon dissolved.

Hence graminivorous birds are provided with gizzards *to break the protecting coats of the grain;* and animals that feed on seeds and nuts *strip them of their shells and husks.*

It is remarkable that in the *succulent fruits,* such as the strawberry, the raspberry, currant, apple, orange, melon, &c., and which, from their very nature, are likely to attract animals to use them, and in eating which *the seeds are likely to be swallowed,* they are fortified by a doubly-protective coating; the pips of the apple, orange, &c., and the seeds of the strawberry and raspberry, pass through the digestive organs, not only unharmed, but their

"And it was commanded them that they should not hurt the grass of the earth, neither any green thing, neither any tree."—REVELATION IX.

germinating powers are even improved by the warmth and trituration of the stomach. Indeed, the stomachs of quadrupeds and birds have been made the vehicles of propagating plants, and distributing them to the widest geographical latitudes. It is even said of some seeds that they will not germinate until they have passed through the digestive ogans of an animal.

1287. *Why do animals that graze, crop the tender blades of grass, but avoid the tall stems?*

Because they are tempted by the greater sweetness and tendernes of the young blades; and in this temptation a very important end is served; for, by avoiding the stems that have grown up, *the animals spare the matured plant by which seeds are borne*, and by which the supply of food is to be continued.

1288. *Why do the eggs of butterflies lie dormant during the winter?*

Because the *coldness of the winter* would be fatal to the life of the young insects; and the absence of vegetation would leave the caterpillars to *perish of starvation*, if they were developed during the winter months.

Fig. 76.—CATERPILLAR FEEDING.

1289. *Why do caterpillars appear in the spring?*

Because the increasing warmth of the sun developes the living embryo, *at the same time that it developes the vegetable germ.* The warmth, therefore, that calls the caterpillar from its embryo sleep, also kindles the germinating power of the vegetable upon which it is destined to feed. The worm awakes and finds the bountiful table of nature spread for it.

"Thou shalt plant vineyards, and dress them, but shalt neither drink of the wine, nor gather the grapes: for the worms shall eat them."—
DEUTERONOMY XXVIII.

1290. *Why does the caterpillar eat voraciously?*

Because it *grows rapidly,* and a large amount of vegetable matter is necessary to supply the rapid growth of its animal substance. Caterpillars in the course of a month devour 60,000 times their own weight of aliment.

Fig. 77.—THE UNDER SIDE OF THE CHRYSALIS OF THE PEACOCK BUTTERFLY.

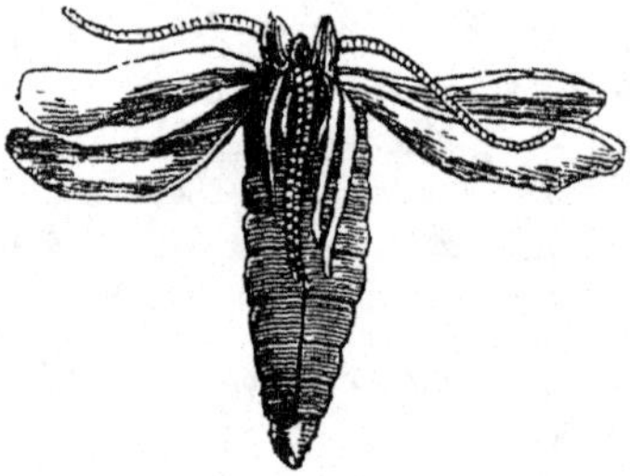

Fig. 78.—THE SAME CHRYSALIS, WITH PART OF ITS SHEATH RAISED TO SHOW THE PARTIALLY-FORMED WINGS, &c.

1291. *Why do caterpillars pass into the state of the chrysalis?*

Because they are thereby prepared for the new existence which they are about to enjoy; *new organs must be perfected in them* to adapt them to the altered conditions of their lives.

Because, also, in the transformation of their bodies, differing materially from the laws of existence that pertain to other creatures, the Creator affords another illustration of his Omnipotence.

Because, also, during the stage that the insect sleeps in the chrysalis, the flowers and their sweet juices, upon which the fly is to feed, are being prepared for it, just as, when it was sleeping in the egg, the green food was being prepared for the caterpillar. When, therefore, the beautiful fly spreads its silken wings, it finds a *second time* that, while it has slept, its meal has been prepared, and it now flies away joyously to feed upon the milk and honey of beautiful flowers which, at the time it passed into the chrysalis, had not yet unfolded their petals.

"For the moth shall eat them up like a garment, and the worm shall eat them like wool. but my righteousness shall be for ever, and my salvation from generation to generation."—ISAIAH LI.

Paley observes, that "the *metamorphosis* of insects from grubs into moths and flies, is an astonishing process. A hairy caterpillar is transformed into a butterfly. Observe the change. We have four beautiful wings where there were none before; a tubular proboscis, in the place of a mouth with jaws and teeth; six long legs, instead of fourteen feet. In another case, we see a white, smooth, soft worm, turned into a black, hard, crustaceous beetle, with gauze wings. These, as I said, are astonishing processes, and must require, as it should seem, a proportionably artificial apparatus.

Fig. 79.—THE PEACOCK BUTTERFLY.

The hypothesis which appears to me most probable, is that, in the grub, there exists at the same time three animals, one within another, all nourished by the same digestion, and by a communicating circulation; but in different stages of maturity. The latest discoveries made by naturalists, seem to favour this supposition. The insect, already equipped with wings, is descried under the membranes both of the worm and nymph. In some species, the proboscis, the antennæ, the limbs, and wings of the fly, have been observed to be folded up within the body of the caterpillar; and with such nicety as to occupy a small space only under the two first wings. This being so, the outermost animal, which, besides its own proper character, serves as an integument to the other two, being the farthest advanced, dies, as we suppose, and drops off first. The second, the pupa or chrysalis, then offers itself to observation. This also, in its turn, dies; its dead and brittle husk falls to pieces, and makes way for the appearance of the fly or moth. Now, if this be the case, or indeed whatever explication be adopted, we have a

"That which the palmer-worm hath left hath the locust eaten; and that which the locust hath left hath the canker-worm eaten; and that which the canker-worm hath left hath the caterpillar eaten."—JOEL I.

prospective contrivance of the most curious kind; we have organisations *three deep;* yet a vascular system, which supplies nutrition, growth, and life, to all of them together."

Lord Brougham, in a note upon the above, does not support Paley's view. He says "It is more than probable that the parts which are to appear in the perfect insect do *not* exist in the larvæ, where there is not much difference between the larva and pupa, excepting at the time just previous to its becoming a pupa, at which time the larva is motionless and torpid. The caterpillar of a moth, when about to turn into a pupa, provides for the protection of the latter state, either by surrounding itself with a web, or by some other means. Soon after this is accomplished, the caterpillar becomes motionless, or nearly so; it can neither eat nor crawl. At this time, and *not before*, the parts of the pupa are forming within the skin of the caterpillar, which may be easily seen by dissection."

It appears to the author, however, that Paley is partially right, and Lord Brougham totally wrong, in these remarks. When Lord Brougham asserts that the parts of the pupa are forming within the skin of the caterpillar at that time when the transformation begins, "and not before, which may be easily seen by dissection," he forgets, that although in some instances it is the first moment when, to the human eye, the organs of the new creature *become perceptible*, that the "*three deep*" nature which Paley attributes to the *grub*, must really have existed *in the egg*—that the *butterfly* originated *in the egg*, as certainly as did the *caterpillar*, or the *crysalis*, and that unless that egg had possessed its three mysterious embryos, it would have been impossible for the grub to have progressed to the stages of transformation. No one has ever known the embryo of a bird's egg to pass through three distinct and dissimilar states of existence; nor has any one ever known the embryo of the butterfly's egg to stop short at either of the stages, if the proper conditions of its existence and development were supplied to it. *Why?* Because the embryo of the insect has a *threefold* nature, while that of the bird is *single.*

"They shall cut down her forest, saith the Lord, though it cannot be searched; because they are more than the grasshoppers, and are innumerable."—
JEREMIAH XLVI.

CHAPTER LXVI.

1292. *Why does the caterpillar become torpid when passing into the state of the chrysalis?*

Because in all probability, where the difference between the first and the ultimate form is considerable, the organs of the insect having to undergo great changes, it would suffer considerable pain. Torpor comes upon the insect, it is thrown into a state similar to that of a person who has inhaled chloroform; and after what has, in all probability, proved a pleasant dream, the insect awakes to find itself changed and beautified.

1293. *Why are the pupæ of grasshoppers and other insects, when about to undergo transformation, still active and sensitive?*

Because, as there is but a *slight difference* between the form which they have in the pupa state, and that which they ultimately assume, they do not require the state of torpidity to save them from pain, nor to arrest their movements while their organs are being changed. With them *the outer skin is thrown off,* and they are then perfect insects.

1294. *Why do caterpillars, when about to pass through the chrysalis state, attach themselves to the leaves of plants, &c.?*

Because they know instinctively that for a time they will be *unable to controul their own movements, and to avoid danger.* They therefore choose secure and dry places, underneath leaves, or in the crevices of old and dry walls, and there they firmly attach themselves, to await the time of their liberation.

1295. *Why do insects attach their eggs, to leaves &c.?*

Because, as the eggs have to be preserved during the winter, the insect attaches them to some surface which will be a *protection to them.* Generally speaking, the eggs are attached to the permanent stems of plants, and not to those leafy portions which are liable to fall and decay. The spider *weaves a silken bag* in which it deposits its eggs, and then it hangs the bag in a sheltered situation. Nature

"Lay up for yourselves treasures in heaven, where neither moth nor rust doth corrupt, and where thieves do not break through and steal."—MATT. VI.

keeps her butterflies, moths, and caterpillars, locked up during the winter, in their egg-state; and we have to admire the various devices to which, if we may so speak, the same nature has resorted for the *security* of the egg. Many insects enclose their eggs in a silken web; others cover them with a coat of hair, torn from their own bodies; some glue them together; and others, like the moth of the silkworm, glue them to the leaves upon which they are deposited, that they may not be shaken off by the wind, or washed away by rain; some again make incisions into leaves, and hide an egg in each incision; whilst some envelope their eggs with a soft substance, which forms the first aliment of the young animal; and some again make a hole in the earth, and, having stored it with a quantity of proper food, deposit their eggs in it.

1296. *Why do butterflies fly by day?*

Because they are *organised to enjoy light and warmth*, and they live upon the sweets of flowers which by day are most accessible.

1297. *Why do moths fly by night?*

Because they are *organised to enjoy subdued light* and cool air; and as they take very little food during the short life they have in the winged state, they find sufficient by night. Some of the moths, like that of the silk-worm, take no food from the time they escape from the chrysalis until they die.

Because, also, they form the food of bats, owls, and other of the night-flying tribes.

1298. *Why are the bodies of moths generally covered with a very thick down?*

Because, as they fly by night, they are liable to the effects of cold and damp. The moths, therefore, are nearly all of them covered with a very thick down, quite distinguishable from the lighter down of butterflies.

1299. *Why do moths fly against the candle flame ?*

Because their eyes are organised *to bear only a small amount of light*. When, therefore, they come within the light of a candle, their sight is overpowered and their vision confused; and as they cannot distinguish objects, they pursue the light itself, and fly against the flame.

"Let him that glorieth glory in this that he understandeth and knoweth me, that I am the Lord which exercise loving-kindness, judgment, and righteousness in the earth: for in these things I delight, saith the Lord."—JER. IX.

1300. *Why do insects multiply so numerously?*

Because they form the food of larger animals, and especially of birds. A single pair of sparrows and a nest of young ones have been estimated to consume upwards of *three thousand* insects in a week.

1301. *Why does the "death-watch" make a ticking noise?*

Because the insect is one of the beetle tribe, having a horny case upon its head, *with which it taps upon any hard substance*, the ticking is the call of the insect to its species, just as the noise made by the cricket is a note of communication with other crickets.

There is a superstition connected with the death-watch, which, like most superstitions, is based upon the theory of *probabilities*. The death-watch is usually heard in the spring of the year, and a superstition runs to the effect that some one in the house will die before the year has ended. Persons who are superstitious are never very strict in the interpretation of their predictions; and therefore, whether a person dies in the house or out of it, in the same room where the death-watch was heard, or across the wide Atlantic, so that there be some kind of relationship, or even acquaintance, between the person who hears the omen, and the person dying, the event is sure to be connected with the prophetic sounds of the death-watch. Little weens the small timber-boring beetle, when he is tapping gently to call his mate, and perhaps peeping into every corner and crevice to find her, that he is sending dismay into the heart of some superstitious listener, who, in ignorance of a simple fact, overwhelms herself with an imaginary grief

1302. *Why are insects in the first stage, after leaving the egg, said to be in the "larva" state?*

Because that name is founded upon the Latin word *larva*, meaning masked, clothed as with a mask; the term is meant to express that the future insect is disguised in its first form.

1303. *Why are insects in the second state said to be in the "pupa" state?*

Because the term is derived from the Latin *pupa*, from a slight resemblance in the manner in which the insects are enclosed, to that in which it was the fashion of the ancients to *bandage their infants*.

1304. *Why are insects in the "pupa" stage also called "chrysalides?"*

Because, as the Latin term implies, it is adorned with gems. Many chrysalides are *studded with golden and pearl-like spots*.

"Thou hast set all the borders of the earth: thou hast made summer and winter."—PSALM LXXIV.

1305. *Why are the perfect insects said to be in the "nymph" state?*

Because their joyful existence, and their beautiful forms, give them a fancied resemblance to the *nymphs of the heathen mythology*. The nymphs were supposed goddesses of the mountains, forests, meadows, and waters.

This term has generally, but very improperly, been also applied to the pupa state, so that *pupa*, *chrysalis*, and *nymph* have all been employed to represent one state. This is obviously an error, as there is nothing in the condition of the *pupa* or *chrysalis* that can at all accord with the mythological idea of a *nymph*, and which, in reference to the beautiful and joyous fly, finds a much truer application.

CHAPTER LXVII.

1306. *Whence does the snail obtain its shell?*

Young snails come from the egg *with a shell upon their backs.*

1307. *How does the shell grow with the increase of size of the animal?*

The soft slime which is yielded by the body of the animal, *hardens upon the orifice of the shell*, and thus increases its size.

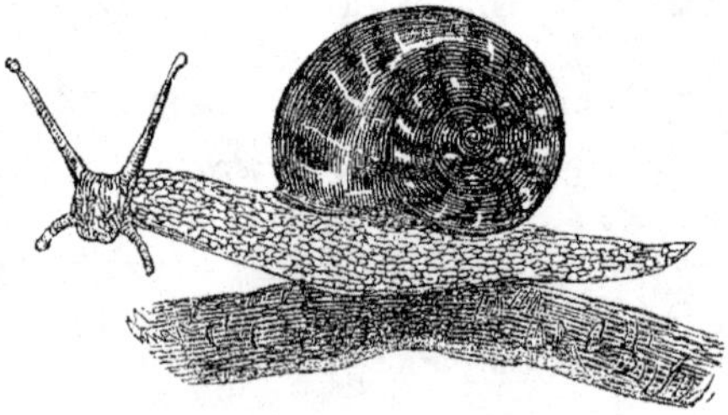

Fig. 80.—COMMON GARDEN SNAIL.

1308. *Why is the shell spiral?*

Partly because of its original formation; but also because, *as the shell grows*, the opening is elongated, and thrown up, causing the spiral body of the shell to turn, and so to wind its growth around the centre.

1309. *Why has the snail four tentacula attached to its head?*

Because the insect, having no other limbs, is provided with those projecting members, the lower two serving as *feelers* and the upper two also as *feelers* and *eyes*. These, projecting in the front of the animal, impart to it a consciousness of surrounding objects, and especially of those which lie in its path.

1310. *Why is the snail able to move, without feet?*

Because it has attached to its body a fringe of muscular skin, which is capable of considerable contraction and expansion, and by alternately stretching and shortening this, the snail is able to draw himself along.

1311. *Why do we see no snails in the winter time?*

Because they bury themselves in the ground, or in holes, where they remain *in a torpid state* for several months. Before they enter into the torpid state, they form with their slimy secretion, and with some earthy matters which they collect, a strong cement with which they seal up the opening to their shells.

1312. *Why can snails live in shells thus sealed?*

Because they leave, in the thin wall by which they close themselves in, *a small hole*, too small to admit water, but large enough to let in sufficient air to carry on their feeble respiration during their winter sleep.

1313. *Why do insects abound in putrid waters, and in decaying substances?*

Because they have been endowed with appetites and with constitutions that enable them to live upon and to enjoy corrupt matter. In this point of view the maggots of flies are exceedingly useful; a dead carcass is speedily threaded by them in every direction; thus that corrupt matter which, in a large mass, would poison the air, is taken up in small portions by millions of living bodies, and by them *dispersed*, and becomes innoxious.

"For he maketh small the drops of water: they pour down rain according to the vapour thereof."—JOB XXXV.

1314. *Why do we see, in tanks of rain water, insects rising to the surface?*

Because numerous insects pass through their first stages of existence in *water*, and among them the common gnat. The gnats of the previous season having deposited their eggs on the sides of the water-butt, the warm water developes them, and the larvæ of the gnats appear (Fig. 81; *c* natural size of larva; *b* larva magnified).

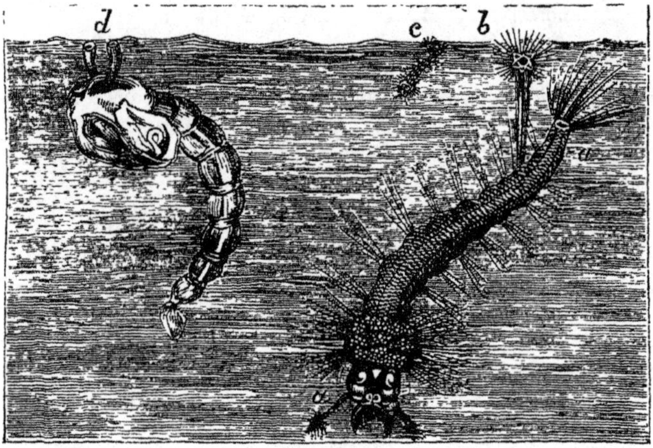

Fig. 81.—LARVA AND PUPA OF GNAT.
(*Greatly magnified.*)

1315. *Why do they continually rise to the surface of the water?*

Because they require to breathe air, and therefore they come up to the surface, where, elevating the tube *(b)* above the surface of the water, they are enabled to breathe.

1316 *Why do some appear to have larger heads than others?*

Those that have apparently larger heads, and that breathe through tubes attached to their heads *(d)* are in the *pupa*, or second stage of development, and underneath the large shield by which their heads are marked, their wings, feet, &c., are being formed.

1317. *Why, when the water is disturbed, do the larvæ descend more rapidly than the pupæ?*

Because the pupæ are in a torpid condition, awaiting the formation of their perfect organs.

1318. *Why are the flies able to escape from the water?*

Because, as their formation becomes perfected, and the fluids of the body of the pupa become absorbed in the production of the light texture of the wings, &c., *the body and its case become lighter than the water*, and rise and float upon the surface. The pupa-case then forms a natural boat, from which the fly emerges, and spreading its wings, enters upon the final state of its existence.

FIG. 82.—THE PERFECT GNAT, ESCAPING FROM THE PUPA-CASE.
(*Greatly magnified.*)

This interesting metamorphosis may be seen going on in the summer time, in every pond, brook, and reservoir. A fine sunny morning calls up millions of these little boats from beneath the surface, and the diver within that wonderful little bell breaks its sealed doors, and flies away to enjoy the bright sunshine.

1319. *Why are beetles denominated "coleoptera?"*

Because they have wings protected by horny sheaths; the term *coleoptera* signifies *wings in a sheath.*

"They shall lie down in the dust; and the worms shall cover them."—JOB XXI.

1320. *Why have beetles hard horny wing-cases?*

Because they live underground, or in holes excavated in wood, &c. If, therefore, their wings were not protected by a hard and firm covering, they would be constantly *liable to destruction* from the movement of the insect within hard and rough bodies.

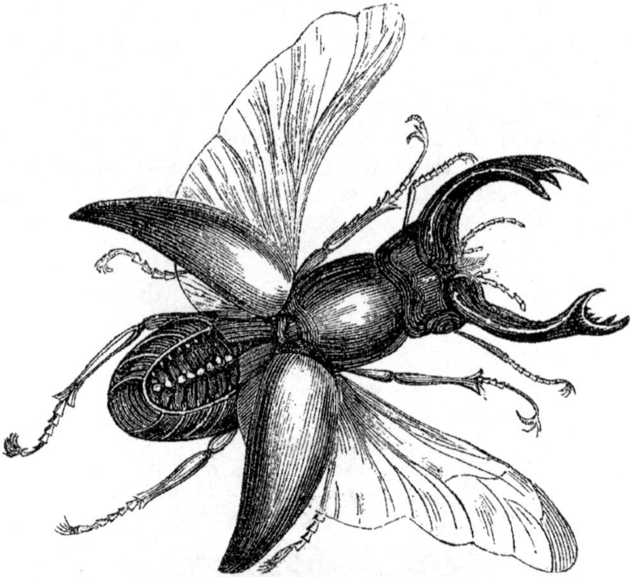

Fig. 83.—STAG-BEETLE, SHOWING ITS WINGS UNFOLDED, AND THE WING CASES OPEN.

The *elytra*, or scaly wings of the genus of scarabæus, or beetle, furnish an example of this kind. The true wing of the animal is a light, transparent membrane, finer than the finest gauze, and not unlike it. It is also, when expanded, in proportion to the size of the animal, very large. In order to protect this delicate structure, and, perhaps, also to preserve it in a due state of suppleness and humidity, a strong, hard case is given to it, in the shape of the horny wing which we call the elytron. When the animal is at rest, the gauze wings lie folded up under this impenetrable shield. When the beetle prepares for flying, he raises the integument, and spreads out his thin membrane to the air. And it cannot be observed without admiration, what a tissue of cordage, *i. e.* of muscular tendons, must run in various and complicated, but determinate directions, along this fine surface, in order to enable the animal, either to gather

"The Lord is good; his mercy is everlasting; and his truth endureth to all generations."—PSALM C.

it up into a certain precise form, whenever it desires to place its wings under the shelter which nature hath given to them, or to expand again their folds when wanted for action.

In some insects, the elytra cover the whole body; in others, half; in others only a small part of it; but in all, they completely hide and cover the true wings. Also,

Many, or most of the beetle species lodge in holes in the earth, environed by hard, rough substances, and have frequently to squeeze their way through narrow passages; in which situation, wings so tender, and so large, could scarcely have escaped injury, without both a firm covering to defend them, and the capacity of folding themselves up under its protection.

1321. *Why have many of the beetle tribe large strong horns?*

Because, as they live in holes in the earth, or in excavations in wood, they use their horns to *dig out their places of retreat.*

1322. *Why has the giraffe a small head?*

Because, being set upon the end of a very long neck, the animal would be *unable to raise it* if it were heavy.

1323. *Why has the giraffe a long neck?*

Because it *feeds upon the branches of tall trees.*

1324. *Why has the giraffe a long and flexible tongue?*

Because it is thereby enabled to lay hold of the tender twigs and branches, *and draw them into its mouth*, avoiding the coarser parts of the branches.

1325. *Why are the nostrils of the giraffe small and narrow, and studded with hairs?*

Because the hairs and the peculiar shape of the nasal passages are designed as a protection against the insects which inhabit the boughs of the trees upon which the giraffe feeds; and also against the sands of the desert, which storms raise into almost suffocating clouds.

1326. The distribution of animals, or *Zoological Geography*, is a of great interest, and should be carefully studied in connection with *Botanical Geography* (*see* 1208). The highest department of the animal kingdom (writes the Rev. W. Milner) commences with the class of *Birds*, which may be naturally divided into the three great orders of aerial, terrestrial, and aquatic. Aggregation into immense flocks is a distinguishing feature of several species, especially of the aquatic order, which form separate colonies, building their nests in the same state, though other spots equally adapted are at no great distance.

Fig. 84.—GIRAFFE FEEDING.

Hence the Vogel-bergs, or bird-rocks of the northern seas, one of which at Westmannsharn in the Faroe group of islands, seldom intruded upon by man, presents a most extraordinary spectacle to the visitor. The Vogel-berg lies in a frightful chasm in the precipitous shores of the island, which rise to the height of a thousand feet, only accessible from the sea by a narrow passage. Here congregate a host of birds. Thousands of guillemots and auks swim in groups around the boat which conveys man to their domain, look curiously at him, and vanish beneath the water to rise in his immediate neighbourhood. The black guillemot comes close to the very oars. The seal stretches his head above the waves, not comprehending what has disturbed the repose of his asylum, while the rapacious skua pursues the puffin and gull. High in the air the birds seem like bees clustering about the rocks, whilst lower they fly past so close that they might be knocked down with a stick. But not less strange is the domicile of this colony. On some low rocks scarcely projecting above the water sit the glossy cormorants, turning their long necks on every side, Next are the skua gulls, regarded with an anxious eye by the kittiwakes above. Nest follows nest in crowded rows along the whole breadth of the rock, and nothing is visible but the heads of the mothers and the white rocks between, A little higher on the narrow shelves sit the guillemots and auks, arranged as on parade, with their white breasts to the sea, and so close that a hailstone could not pass between them. The puffins take the highest station, and, though scarcely visible, betray themselves by their flying backwards and forwards. The noise of such a multitude of birds is confounding, and in vain a person asks a question of his nearest neighbour. The harsh tones of the kittiwakes are heard above the whole, the intervals being filled with the monotonous note of the auk, and the softer voice of the guillemot. When Graba, from whose travels this description is principally drawn, visited the Vogel-berg, he was tempted by the sight of a crested cormorant to fire a gun, but what became of it, he remarks, it was impossible to ascertain. The air was darkened by the birds roused from their repose. Thousands hastened out of the chasm with a frightful noise, and spread themselves over the ocean. The puffins came wandering from their holes, and regarded the universal confusion with comic gestures. The kittiwakes remained composedly in their nests, whilst the cormorants tumbled headlong into the sea. Similar great congregations of the feathered race appear where the shores are rocky high, and precipitous, but this is strikingly the case, where

—"The northern ocean, in vast whirls,
Boils round the naked melancholy isles
Of farthest Thule; and the Atlantic surge
Pours in among the stormy Hebrides.
Who can recount what transmigrations there
Are annual made? what nations come and go?
And how the living clouds on clouds arise?
Infinite wings! till all the plume-dark air
And rude resounding shore are one wild cry."

1327. Most terrestrial birds, unacquainted with man, exhibit a remarkable tameness, and are slow in acquiring a dread of him, even after repeated lessons that danger is to be apprehended from his neighbourhood. Mr. Darwin speaks of a gun as almost superfluous in the unfrequented districts of South America, for with its muzzle he pushed a hawk off the branch of a tree. Once, while lying down, a mocking thrush alighted on the edge of a pitcher, made of the shell of a tortoise, which he was holding in his hand, and began very leisurely to sip the water, even allowing him to handle it while seated on the vessel. In Charles Island, which had been colonised about six years, he saw a boy sitting by a well with a switch in his hand, with which he killed the doves and finches as they came to drink; and for some time had been constantly in the habit of waiting by the well for the same purpose, to provide himself with his dinners. In the Falkland Islands, at Bourbon, and at Tristan d'Acunha, the same tame-

"As a bird that wandereth from her nest; so is a man that wandereth from his place."—PSALM XXVII.

ness was noticed by the early visitors. On the other hand, the small birds in the arctic regions of America, which have never been persecuted, exhibit the anomalous fact of great wildness. From a review of various facts, Mr. Darwin concludes, "first, that the wildness of birds with regard to man is a particular instinct directed against him, and not dependent on any general degree of caution arising from other sources of danger; secondly, that it is not acquired by individual birds in a short time, even when much persecuted; but that in the course of successive generations it becomes hereditary. Comparatively few young birds in any one year have been injured by man in England, yet almost all, even nestlings, are afraid of him; many individuals, however, both at the Galapagos and at the Falklands, have been pursued and injured by man, but yet have not learned a salutary dread of him."

1328. **Numerous species of birds may be regarded as the favourites of nature on account of the gracefulness given to their shape**, and the richly-coloured **plumage with which they are adorned**, as evidenced in the gaudy liveries of **many of the parrot tribe**, and the forms and hues of the birds of paradise. But **they are especially** interesting to man for the faculty of song with which they **are endowed**, in some, "most musical, most melancholy," in others, sprightly and animating, inspiriting the sons of toil under the burdens peculiar to their station. It deserves to be remarked, as an instance of compensation and adjustment, that whilst the birds of the temperate zone are far inferior to those of tropical climes in point of beauty, they have far more melodious notes in connection with their less attractive appearance.

1329 From the powerful means of locomotion possessed by several of the bird tribe, and their great specific levity, air being admitted to the whole organisation as water to a sponge, it might be inferred, that the entire atmosphere was intended to be their domain, so that no species would be limited to a particular region. The common crow flies at the rate of twenty-five miles an hour; the rapidity of the eider-duck, *Anas mollissima*, is equal to ninety miles an hour; while the swifts and hawks travel at the astonishing speed of a hundred and fifty miles in the same time. It is true that some species have a very extensive range, as the nightingale, the common wild goose, and several of the vulture tribe. The same kind of osprey or fishing-eagle that wanders along the Scottish shores appears upon those of the south of Europe, and of New Holland. The lammergeyer haunts the heights of the Pyrenees, the mountains of Abyssinia, and the Mongolian steppes; and the penguin falcon occurs in Greenland, Europe, America, and Australia. In general, however, like plants and terrestrial quadrupeds, the birds are subject to geographical laws, definite limits circumscribing particular groups. The common grouse of our own country affords a striking exemplification of this arrangement, as it is nowhere met with out of Great Britain; and other examples occur of a very scanty area containing a species not to be found in any other region. The celebrated birds of paradise are exclusively confined to a small part of the torrid zone, embracing New Guinea and the contiguous islands; and the beautiful Lories are inhabitants of the same districts, being quite unknown to the New World. Parroquets are chiefly occupants of a zone extending a few degrees beyond each tropic, but the American group is quite distinct from the African, and neither of these have one in common with the parrots of India, The great eagle is limited to the highest summits of the Alps; and the condor, which soars above the peak of the loftiest of the Andes, never quits that chain. Humming-birds are entirely limited to the western hemisphere, where a particular species is sometimes bounded by the range of an island, while others are more extensively spread, the *Trochilus flammifrons*, common to Lima, being observed by Captain King upon the coast of the Straits of Magellan, in the depth of winter, sucking the flowers of a large fuschia, then in bloom in the midst of a shower of snow. Among the birds incapable of flight, which rival the quadrupeds in their size, the intertropical countries of the globe have their distinct species, presenting similar general features of organisation, as the ostrich of Africa and Arabia, the cassowary of Java and Australia, and the touyou of Brazil. In the arctic regions, we meet with species peculiar to them, the *Strix laeponicus* or Lapland owl, and the eider-duck, an inhabitant of the shores, from whose nests the eider-down is obtained. Several families of maritime birds are likewise limited to particular oceanic localities. Approaching the fortieth parallel of latitude, the albatross is seen flitting along the surface of the waves, and soon afterwards the frigate and other tropical birds appear, which never wander far beyond the torrid zone. It thus appears, that, notwithstanding the great locomotive powers of birds, particular groups have had certain regions assigned to them as their sphere of existence, which they are adapted to occupy, and to which they adhere in the main, though it is easy to conceive of natural causes occasionally constraining to a migration into new and even distant territories. Captain Smyth informed Mr. Lyell, that when engaged in his survey of the Mediterranean, he encountered a gale in the Gulf of Lyons, at the distance of between twenty and thirty leagues from the coast of France, which bore along many land-birds of various species, some of which alighted on the ship, while others were thrown with violence against the sails. In this manner, many an islet in the deep, after ages of solitude and silence, uninterrupted except by the wave's wild dash, and the wind's fierce howl, may have received the song of birds, forced by the tempest from their home, and compelled to seek a new one under its direction.

1330. There is no feature more remarkable in the economy of birds than the periodical migrations, so systematically conducted, in which five-sixths of the whole feathered population engage. In the case of North America, according to an estimate by Dr. Richardson, the passenger-pigeons form themselves into vast flocks for the journey, one of which has been calculated to include 2,230,000,000 individuals. We are familiar with the cuckoo as our visitor in spring, and with the house-swallow as our guest through the summer, the latter usually departing in October to the warmer regions of the south, wintering in Africa, returning again when a more genial season revives its insect food. By cutting off two claws from the feet of a certain number of swallows, Dr. Jenner ascertained the fact of the same individuals re-appearing in their old haunts in the following year, and one was met with even after the lapse of seven years. The arctic birds migrate farther south, when the seas, lakes, and rivers become covered with unbroken sheets of ice; the swans, geese, ducks, divers, and coots flying off in regular phalanxes to regions where a less rigorous winter allows of access to the means of life. Hence, soon after, we lose the swallows, we gain the snipes and other waders, which have fled from the hard frozen north to our partially frozen morasses, where their ordinary nutriment may still be obtained. The equinoctial zone, where the seasonal change is that of humidity and drought furnishes an example of the same phenomenon. As soon as the Orinoco is swollen by the rains, overflows its banks, and inundates the country on either

"The Lord is my light and my salvation; whom shall I fear? the Lord is the strength of my life; of whom shall I be afraid?"—PSALM XXVII.

side, an innumerable quantity of aquatics leave its course for the West India islands on the north, and the valley of the Amazon on the south, the increased depth of the river, and the flooded state of the shores, depriving them of the usual supply of fish and insects. Upon the stream decreasing, and retiring within its bed, the birds return.

1331. A comparison between the quadrupeds of the Old and New Worlds is in every point strikingly in favour of the former. Not only has the western continent no animals of such giant bulk as those of the eastern, but no examples of such high organisation, such power and courage, as the African lion and the Asiatic tiger display. Buffon's remark must indeed be considerably modified, respecting the cowardice of the American feline race, for the jaguar of the woods about the Amazon, when attacked by man, will not hesitate to accept his challenge, will even become the assailant, nor shrink from an encounter against the greatest odds. The following passages from the writings of Humboldt show that this transatlantic animal is not to be despised:—

"The night was gloomy; the Devil's Wall and its denticulated rocks appeared from time to time at a distance, illuminated by the burning of the savannahs, or wrapped in ruddy smoke. At the spot where the bushes were the thickest, our horses were frightened by the yell of an animal that seemed to follow us closely. It was a large jaguar, that had roamed for three years among these mountains. He had constantly escaped the pursuit of the boldest hunters, and had carried off horses and mules from the midst of enclosures; but, having no want of food, had *not yet* attacked men. The negro who conducted us uttered wild cries. He thought he should frighten the jaguar; but these means were of course without effect. The jaguar, like the wolf of Europe, follows travellers even when he will not attack them; the wolf in the open fields and in unsheltered places, the jaguar skirting the road, and appearing only at intervals between the bushes."

The same illustrious observer also remarks,—

"Near the Joval, nature assumes an awful and savage aspect. We there saw the largest jaguar we had ever met with. The natives themselves were astonished at its prodigious length, which surpassed that of all the tigers of India I had seen in the collections of Europe."

Still these were extraordinary specimens of the race, and leave the fact undoubted, that the most formidable of the western *Feræ* has no pretensions to an equality with his congener, the tyrant of the jungles of Bengal.

1332. In vain also we look among the tribes of America for a rival in outward appearance to the giraffe, so remarkable for its height, its swan-like neck, gentle habits, and soft expressive eye: while of the animals most serviceable to mankind—the horse, the ox, the ass, the goat, and the hog—not a living example of either was known there before its occupancy by the Europeans. But, however inferior the animal race of the New may be as compared to those of the Old world, the balance between the two appears to have been pretty equal in remote ages; geological discovery has disproved the assertion of Buffon, that the creative force in America in relation to quadrupeds never possessed great vigour, and has established the fact, that it is only the more recent specimens of its energy that are upon an inferior scale. The relics of the unwieldly magatherium, of the gigantic sloth, and armadillo-like animals, discovered in great abundance imbedded in its soil, prove that at a former period it swarmed with monsters of equal bulk with those that now roam in the midst of Africa and Asia. The estuary deposit that forms the plains westward of Buenos Ayres, and covers the gigantic rocks of the Bando Oriental, appears to be the grave of extinct gigantic quadrupeds.

"But wild beasts of the desert shall lie there; and their houses shall be full of doleful creatures; and owls shall dwell there, and satyrs shall dance there."—ISAIAH XIV.

1333. There are various animals which are very widely dispersed, enduring the extremes of tropical heat and of polar cold, which are either in a wild condition or in a state of domestication. Wild races, considered to be varieties of the domestic dog, occur in India, Sumatra, Australia, Beloochistan, Natolia, Nubia, various parts of Africa, and both the Americas; while in subjection to man, the dog is his faithful companion, and has followed his steps into every diversity of climate and of situation to which he has wandered. The north temperate zone of the Old Continent appears to be the native region of the ox, which passes in Lapland within the arctic circle, and has been spread over South America since its first introduction by the Spaniards. The horse, originally an inhabitant of the temperate parts of the Old World, has shared in a similar dispersion, and now exists in the high latitude of Iceland, in the desolate regions of Patagonia, and roams wild in immense herds over the Llanos of the Orinoco, leading a painful and restless life in the burning climate of the tropics. Humboldt draws a striking picture of the sufferings of these gifts of the Old World to the New, returned to a savage state in their western location.

"In the rainy season, the horses that wander in the savannah, and have not time to reach the rising grounds of the Llanos, perish by hundreds amidst the overflowings of the rivers. The mares are seen, followed by their colts, swimming, during a part of the day, to feed upon grass, the tops of which alone wave above the waters. In this state they are pursued by the crocodiles; and it is by no means uncommon to find the prints of the teeth of these carnivorous reptiles on their thighs. Pressed alternately by excess of drought and of humidity, they sometimes seek a pool, in the midst of a bare and dusty soil, to quench their thirst; and at other times flee from water and the overflowing rivers, as menaced by an enemy that encounters them in every direction. Harassed during the day by gad-flies and mosquitoes, the horses, mules, and cows find themselves attacked at night by enormous bats, that fasten on their backs, and cause wounds which become dangerous, because they are filled with acaridæ and other hurtful insects. In the time of great drought, the mules gnaw even the thorny melocactus (melon-thistle), in order to drink its cooling juice, and draw it forth as from a vegetable fountain. During the great inundations, these same animals lead an amphibious life, surrounded by crocodiles, water-serpents, and manatees Yet, such are the immutable laws of nature, their races are preserved in the struggle with the elements, and amid so many sufferings and dangers. When the waters retire, and the rivers return into their beds, the Savannah is spread over with a fine odoriferous grass, and the animals of old Europe and Upper Asia seem to enjoy, as in their native climates the renewed vegetation of spring."

1334. The first colonists of La Plata landed with seventy-two horses, in the year 1535, when, owing to a temporary desertion of the colony, the animals ran wild; and in 1580, only forty-five years afterwards, it had reached the Straits of Magellan. The ass has a more restricted range than the horse, not being capable of enduring so great a degree of cold, though usually far from being considered a delicate animal. To the warmer parts of the temperate zone, between the 20th and the 40th parallels of latitude, the ass seems best adapted, not propagating much beyond the 60th, and only occurring in a state of degeneration beyond the 52nd. The sheep and goat tribe are widely spread, equally supporting the extremes of temperature. According to Zimmerman, the *Argali* or *Mouflon*, the original race of sheep, still exists on all the great mountains of the two continents; and the *Capricorn* and *Ibex*, the ancestors of the common goat inhabit the high European elevations. From the 64th degree of north latitude the hog is met with all over the old continent, and also in the islands of the Indian Ocean, peopled by the Malay race; and since its introduction into the

"His going forth is from the end of the heaven, and his circuit unto the ends of it, and there is nothing hid from the heat thereof."—PSALM XIX.

New World, it has diffused itself over it, from the 50th parallel north as far as Patagonia. Originally the cat was not known in America, nor in any part of Oceanica, but it has now spread into almost every country of the globe. Among animals entirely wild, the most extensively diffused, are the fox, hare, squirrel, and ermine; but the species are different in every region of the world; nor is there perhaps one example to be found of a species perfectly identical naturally existing in distant localities of the earth.

Respecting the *internal constitution and heat of the earth,* differences of opinion, and some very wild speculation have existed. We find in Humboldt's "Cosmos" the following remarks:—

1335. "It has been computed at what depths liquid and even gaseous substances, from the pressure of their own superimposed strata, would attain a density exceeding that of platinum, or of iridium; and in order to bring the actual degree of ellipticity, which was known within very narrow limits, into harmony with the hypothesis of the infinite compressibility of matter, Leslie conceived the interior of the Earth to be a hollow sphere, filled with "an imponderable fluid of enormous expansive force." Such rash and arbitrary conjectures have given rise, in wholly unscientific circles, to still more fantastic notions. The hollow sphere has been peopled with plants and animals, on which two small subterranean revolving planets, Pluto and Proserpine, were supposed to shed a mild light. A constantly uniform temperature is supposed to prevail in these inner regions, and the air being rendered self-luminous by compression, might well render the planets of this lower world unnecessary. Near the north pole, in 82 deg. of latitude, an enormous opening is imagined, from which the polar light visible in Aurora streams forth, and by which a descent into the hollow sphere may be made. Sir Humphry Davy and myself were repeatedly and publicly invited by Captain Symmes to undertake this subterranean expedition, so powerful is the morbid inclination of men to fill unseen spaces with shapes of wonder, regardless of the counter-evidence of well-established facts, or universally recognised natural laws. Even the celebrated Halley, at the end of the 17th century, hollowed out the earth in his magnetic speculations; a freely rotating subterranean nucleus was supposed to occasion, by its varying positions, the diurnal and annual changes of the magnetic declination. It has been attempted in our own day, in tedious earnest, to invest with a scientific garb that which, in the pages of the ingenious Holberg, was an amusing fiction."

The following are among the speculations which Humboldt thus severely but justly condemns:—

"The increase of temperature observed is about 1 deg. Fahr. for every fifteen yards of descent. In all probability, however, the increase will be found to be in a geometrical progression as investigation is extended; in which case the present crust will be found to be much thinner than we have calculated it to be. And should this be found to be correct, the igneous theory will become a subject of much more importance, in a geological point of view, than we are at present disposed to consider it. Taking, then, as correct, the present observed rate of increase, the temperature would be as follows:

Water will boil at the depth of 2,430 yards.
Lead melts at the depth of 8,400 yards.
There is red heat at the depth of 7 miles.
Gold melts at 21 miles.
Cast iron at 74 miles.
Soft iron at 97 miles.

And at the depth of 100 miles there is a temperature equal to the greatest artificial heat yet observed, a temperature capable of fusing platina, porcelain, and indeed every refractory substance we are acquainted with. These tempera tures are calculated from Guyton Morveau's corrected scale of Wedgwood's pyrometer; and if we adopt them, we find that the earth is fluid at the depth

"He hath filled the hungry with good things; and the rich he hath sent empty away."—LUKE I.

of 100 miles from the surface, and that even in its present state very little more than the soil on which we tread is fit for the habitation of organised beings."

Robert Were Fox, Esq., made a careful enquiry into the theory in question, and by ascertaining the temperatures of Cornish mines at various depths, greatly assisted the solution of a difficult problem.

Upon this subject Mr. Hunt, in his "Poetry of Science," says:—

1336. "A question of great interest, in a scientific point of view, is the temperature of the centre of the earth. We are, of course, without the means of solving this problem, but we advance a little way onwards in the inquiry by a careful examination of subterranean temperature at such depths as the enterprise of man enables us to reach. These researches show us, that where the mean temperature of the climate is 50 deg., the temperature of the rock at 59 fathoms from the surface is 60 deg., at 132 fathoms it is 70 deg; at 239 fathoms it is 80 deg.; being an increase of 10 deg. at 59 fathoms deep, or 1 deg. in 35.4 feet; of 10 deg. more at 73 fathoms deeper, or 1 deg. in 43.8 feet; and of 10 deg. more at 114 fathoms still deeper, or 1 deg. in 64 2 feet.

Although this would indicate an increase to a certain depth of about one degree in every fifty feet, yet it would appear that the rate of increase diminishes with the depth. It appears therefore probable, that the heat of the earth, so far as man can examine it, is due to the absorption of the solar rays by the surface. The evidences of intense igneous action at a great depth cannot be denied, but the doctrine of a cooling mass, and of the existence of an incandescent mass, at the earth's centre, remains but one of those guesses which active minds delight in."

Upon the subject of *hunger* and *thirst*, by which living creatures are prompted to feast upon the bounties of nature, Sir Charles Bell says, in "Appendix to Paley's Natural Theology:"—

1337. "Hunger is defined to be a peculiar sensation experienced in the stomach from a deficiency of food. Such a definition does not greatly differ from the notions of those who referred the sense of hunger to the mechanical action of the surfaces of the stomach upon each other, or to a threatening of chemical action of the gastric juice on the stomach itself. But an empty stomach does not cause hunger. On the contrary, the time when the meal has passed the stomach is the best suited for exercise, and when there is the greatest alacrity of spirits The beast of prey feeds at long intervals; the snake and other cold-blooded animals take food after intervals of days or weeks. A horse, on the contrary, is always feeding. His stomach, at most, contains about four gallons, yet throw before him a truss of tares or lucerne, and he will eat continually. The emptying of the stomach cannot, therefore, be the cause of hunger.

"The natural appetite is a sensation related to the general condition of the system, and not simply referable to the state of the stomach; neither to its action, nor its emptiness, nor the acidity of its contents; nor in a starved creature will a full stomach satisfy the desire of food. Under the same impulse which makes us swallow, the ruminating animal draws the morsel from its own stomach.

1338. "Hunger is well illustrated by thirst. Suppose we take the definition of thirst—that it is a sense of dryness and constriction in the back part of the mouth and fauces; the moistening of these parts will not allay thirst after much fatigue or during fever. In making a long speech, if a man's mouth is parched, and the dryness is merely from speaking, it will be relieved by moistening, but if it comes from the feverish anxiety and excitement attending a public exhibition, his thirst will not be so removed. The question, as it regards thirst, was brought to a demonstration by the following circumstance. A man having a wound low down in his throat, was tortured with thirst: but no quantity of fluid passing through his mouth and gullet, and escaping by the wound, was found in any degree to quench his thirst.

"Thirst, then, like hunger, has relation to the general condition of the animal system—to the necessity for fluid in the circulation. For this reason, a man dying from loss of blood suffers under intolerable thirst. In both thirst and hunger, the supply is obtained through the gratification of an appetite; and as to these appetites, it will be acknowledged that the pleasures resulting from them far exceed the pains. They gently solicit for the wants of the body; they are the perpetual motive and spring to action."

Our task draws near to a conclusion; and we hope that those who have followed our teachings will thirst after further knowledge; that they will henceforward regard the great Book of Nature as the work of an Almighty Hand, and endeavour to find, for everything that Nature does, the *Reason Why*.

A high perception of the wisdom of the Divine Being, must necessarily be the result of an intelligent contemplation of the Divine works. To the ignorant, the name of God is an unmeaning word; it may inspire fear, but it does not develope love. To the dark mind of the untaught man, God is no more than one of those mysterious existences that awe the superstitious, and deter the wicked. There is no grafting of the soul of the man upon the eternal love. But knowledge brings man into communion with that Almighty wisdom which is the fountain of all truth and happiness. To the enlightened man, God is the sun of all goodness, around whom the attributes of Power, Wisdom, and Love, radiate and fill the universe. As man's physical eye cannot withstand the light of the sun, neither can man's spiritual eye see the whole glory of God. But as we can rejoice in the sunshine, and interpret the mission of the sunbeam, so can we find happiness in the Divine presence, and gather wisdom by the contemplation of the Creator's works.

Nature is a great teacher. What a lesson may be gathered from the germination of a seed; how uniformly the germs obey their destiny. However carelessly a seed may be set in the ground, the germ which forms the root, and that which is the architect of the stem, will seek their way—the one to light, the other to darkness—to fulfil their duty. The obstruction of granite rocks, cannot force the rootlet upward, nor drive the leaflet down. They may kill the germs by exhausting their vital powers in an endeavour to find the proper elements; but no obstruction can make a single blade of grass do aught but strive to fulfil the end for which it was created. Would that man were equally true to the purpose of his existence, and suffered neither the rocks of selfishness, nor the false light of temptation, to force or allure him from duty to his God.

J. & W. Rider, Printers, Bartholomew Close, London.

Printed in Dunstable, United Kingdom

91219506R00201